3-12-20

P7-EPC-699

D0122661

The Madhouse Effect

The Madhouse Effect

How Climate Change Denial Is
Threatening Our Planet, Destroying Our Politics,
and Driving Us Crazy

MICHAEL E. MANN and TOM TOLES

Columbia University Press
New York

Columbia University Press
Publishers Since 1893
New York   Chichester, West Sussex
cup.columbia.edu

Copyright © 2016 Michael E. Mann and Tom Toles
All rights reserved

Library of Congress Cataloging-in-Publication Data
Names: Mann, Michael E., 1965– | Toles, Tom.
Title: The madhouse effect : how climate change denial is threatening our
planet, destroying our politics, and driving us crazy / Michael E. Mann
and Tom Toles.
Other titles: Climate change denial is threatening our planet, destroying our
politics, and driving us crazy
Description: New York : Columbia University Press, [2016] |
Includes bibliographical references and index.
Identifiers: LCCN 2016009004 (print) | LCCN 2016010170 (ebook) |
ISBN 9780231177863 (cloth : alk. paper) | ISBN 9780231541817 (e-book)
Subjects: LCSH: Climatic changes. | Global warming. | Climatic
changes—Psychological aspects. | Global warming—Psychological aspects. |
Denial (Psychology)
Classification: LCC QC903 .M3625 2016 (print) | LCC QC903 (ebook) |
DDC 363.738/74—dc23
LC record available at http://lccn.loc.gov/2016009004

Columbia University Press books are printed on permanent and durable acid-free paper.
Printed in the United States of America

c 10 9 8 7 6 5 4 3 2 1

COVER IMAGE: Tom Toles

Michael Mann dedicates this book to his wife, Lorraine Santy,
and daughter, Megan Dorothy Mann,
and to the memory of his brother Jonathan Clifford Mann.

Tom Toles dedicates this book to his wife, Gretchen;
his daughter, Amanda;
and his son, Seth.

CONTENTS

Coauthors Tom Toles (*left*) and Michael Mann (*right*) with their preferred implements.

PREFACE

Why We Wrote This Book

What would bring a pointy-headed, lab-coat wearing, left-brained scientist and a laid-back, artistic, right-brained editorial cartoonist-satirist to collaborate on a book?

The answer is simple: *climate change.*

The warming of the planet caused by our profligate burning of fossil fuels poses perhaps the greatest challenge that human civilization has yet faced. But thus far, we have not met the challenge. We have failed to engage in the actions necessary to prevent dangerous and possibly irreversible changes in our climate.

Why? It's not as if the science isn't compelling. It is.

It's not as if the threat isn't clear. It is.

It's not as if many smart, informed, and concerned individuals haven't sought to bring attention to this crisis. They have.

It isn't even that the overwhelming majority of citizens don't recognize the urgency of acting on the problem. They do.

So how is that we have arrived in a madhouse atmosphere where politicians are able to do the bidding of powerful fossil fuel interests while ignoring the long-term good of the people they are supposed to represent?

Both of us have asked ourselves this question many times.

A scientist tries to understand the way the world works. An editorial cartoonist tries to show the ways it doesn't.

These jobs exist in happy parallel universes until something very unusual happens to bring them together. In this case, that something unusual is the outrageous distortion of science in public policy. The distortion, denial, and confusion in the public-policy response to climate change has been nothing short of a madhouse. The challenge has been, and still is, to figure out where we are and find the way out.

And this is where the scientist and the cartoonist meet. The scientist tries to understand what the facts and implications are. The cartoonist reads about the growing dangers and tries to vividly illustrate them.

Each of us in his own way is passionate about communicating the urgency of the problem—Mike through public lectures, media interviews, and the university courses that he teaches and Tom through his famous editorial cartoons and blogs in the *Washington Post*, which have used humor and satire to highlight this issue for millions of readers.

It occurred to both of us that conventional approaches are not succeeding—at least not quickly enough to avert a catastrophe. Perhaps we need to talk and think about the problem in a different way.

Maybe, we thought, we need to combine our left brains and our right brains, use both our cerebral cortices and our hearts.

And thus was born this collaboration between an editorial cartoonist known for his biting sociopolitical commentary and a climate researcher known largely for his scientific work.

Despite our very different vocations, both of us are committed to informing the public discourse over what is arguably the greatest challenge yet faced by human civilization: the challenge to avert catastrophic interference with Earth's climate.

There is a fire in the house, almost a literal one. But even as the evidence has become unmistakable, and even though the alarm has been sounded several times, public policy has been paralyzed—sometimes from ignorance, sometimes from uncertainty, but often from a campaign of deliberate misinformation.

This is the madhouse of the climate debate. We have followed Alice through the looking glass. White roses here are painted red, and words suddenly mean something different from what they used to mean. The very language of science itself, of "skepticism" and "evidence," is used in a way opposite of how science really employs it.

Not everyone wants the facts to be known. We have run squarely into what Upton Sinclair famously warned us about: "It is difficult to get a man to understand something, when his salary depends on his not understanding it." And there are many powerful interests whose salary has very much depended on the public not understanding climate science.

And so the battle continues to get the world to recognize the possibly permanent damage we are doing to Earth, to ourselves, and to every ecosystem. The fight to preserve the forests, lakes, mountains, even oceans of the only planet we have is not like other battles. This one will not be ending. Time is no longer on our side. Time will not be healing wounds but instead will be inflicting more and more of them if we don't stop shambling around like confused zombies.

In this book, we talk about the basic science underlying climate change, of course, because, frankly, you can't solve a problem if you don't understand what it is in the first place. But first we talk about *science* itself because you have to know what it *is* before you can understand what it *is not*. In the debate over climate change, pseudoscience or antiscience is too often allowed to masquerade as science, and denialism is allowed to pose as skepticism.

The media too often throw up their collective hands and say, "We can't tell the difference." So we get the problem of "false balance," wherein industry propaganda too often is given an equal place on the media stage with actual science when it comes to the issue of climate change. And the rest of us pay the price in the form of a skewed public discourse. This is a problem that one of us has explored in his cartoons and the other has discussed in his previous writings—and we explore it together in depth in this book.

The basic facts are now clear and essentially beyond dispute.

It is time to put out the fire. From presidents to prime ministers to the pope, people are finally waking up to the reality and the challenge. We have seen glimpses of this new awakening—in the record crowds that have marched through the streets of New York City and in cities around the world demanding climate action, in the unprecedented commitments made by the nations of the world to reduce their carbon emissions at the historic Paris climate summit in 2015. But every time in the past that we have started to realize we needed to act, the purveyors of confusion and denial have stepped in to slow us down and lead us astray.

This time we need to stay the course and get it right. We are out of extra chances.

We explore the reasons why we all *should* care about climate change, and we draw attention to the absurd efforts by special interests and partisan political figures to confuse the public, attack the science and the scientists themselves, and deny that a problem even exists. Despite the late hour and the monumental challenge, we believe there is still time and still hope.

In the end, we hope that our readers will learn something, be entertained and outraged, but also be inspired and motivated. Make no mistake about it. We control our own destiny, and we are poised on a precipice. If we continue with the course we are on, our destiny may indeed be to leave behind an unlivable planet of destroyed ecosystems and continuous, unpredictable chaos. But there is another potential future that lies ahead, a future where we embrace the goal of a sustainable planetary existence and take the steps now to ensure that we preserve the health of our planet for generations to come. It is truly up to us.

It is time finally to escape the madhouse.

ACKNOWLEDGMENTS

Michael Mann thanks the many individuals who have provided help and support over the years. First and foremost are his family: his wife, Lorraine; daughter, Megan; parents, Larry and Paula; brothers, Jay and Jonathan; and the rest of the Manns, Sonsteins, Finesods, and Santys.

Mann gives special thanks to Bill Nye, "the Science Guy," for his friendship, leadership, and support over the years.

Mann is greatly indebted to the various policymakers on both sides of the aisle who stood up against powerful interests to support and defend him and other scientists against politically motivated attacks and who have worked to advance the cause of an informed climate policy discourse. Among them are Sherwood Boehlert, Jerry Brown, Bill Clinton, Hillary Clinton, Al Gore, Mark Herring, Bob Inglis, Jay Inslee, Edward Markey, Terry McAuliffe, John McCain, Jim Moran, Harry Reid, Arnold Schwarzenegger, Henry Waxman, Sheldon Whitehouse, and their various staff.

Mann thanks various colleagues in the Departments of Meteorology, Geosciences, and elsewhere at Penn State for the supportive environment they have helped foster. Among them are President Eric Barron, former president Graham Spanier, Dean Bill Easterling, Institutes of Energy and the Environment director Tom Richard, Earth and Environmental

Systems Institute director Sue Brantley, Department of Meteorology head David Stensrud, and former department head Bill Brune.

Mann also wishes to thank the various other friends, supporters, and colleagues past and present for their assistance, collaboration, and friendship over the years, including John Abraham, Ken Alex, Richard Alley, Ray Bradley, Nick Carpino, Kim Cobb, Ford Cochran, John Cook, Leila Conners, Jason Cronk, Jen Cronk, Heidi Cullen, Fred Damon, Kert Davies, Dider de Fontaine, Brendan Demelle, Steve D'Hondt, Henry Diaz, Leondaro DiCaprio, Paulo D'Oderico, Pete Dominick, Kerry Emanuel, Howie Epstein, Jenni Evans, Pete Fontaine, Peter Frumhoff, Jose Fuentes, Nellie Gorbea, David Graves, Thom Hartmann, Tony Haymet, Ben Horton, Malcolm Hughes, Jan Jarrett, Phil Jones, Jim Kasting, Bill Keene, Kalee Kreider, Lee Kump, Deb Lawrence, Scott Mandia, John Mashey, Roger McConchie, Bill McKibben, Pete Meyers, Sonya Miller, Chris Mooney, Ray Najjar, Gerald North, Michael Oppenheimer, Naomi Oreskes, Tim Osborn, Jonathan Overpeck, Rajendra Pachauri, Rick Piltz, Stefan Rahmstorf, Cliff Rechtschaffen, Andra Reed, Catherine Reilly, Scott Rutherford, Barry Saltzman, Ben Santer, Gavin Schmidt, Steve Schneider, Eugenie Scott, Drew Shindell, Hank Shugart, David Silbert, Peter Sinclair, Dave Smith, Jodi Solomon, Richard Somerville, Amanda Staudt, Eric Steig, Byron Steinman, Sean Sublette, Larry Tanner, Lonnie Thompson, Kim Tingley, Dave Titley, Kevin Trenberth, Fred Treyz, Ana Unruh-Cohen, Dave Verardo, David Vladeck, Nikki Vo, Bud Ward, Ray Weymann, John B. Williams.

Tom Toles, evidently being a monumentally less grateful person than his coauthor, has a scanty list of thanks. He would like to thank in particular Fred Hiatt, editorial page editor of the *Washington Post*, for his patience and support of the many, many climate cartoons and blog posts that Toles has produced in his fourteen years at the paper.

Toles would also like to thank everyone anywhere who has made any effort to advance understanding or action on the climate issue. It has rightly been described as the problem from hell because of the difficulties associated with describing and dealing with it in an adversarial political system. It has been a slow, frequently thankless journey for everyone every step of the way, and there are many miles still to travel.

Toles would like specifically to thank wholeheartedly his wife, Gretchen; his children, Amanda and Seth; his brother, George; and all his extended family for all the support and love they have shown over the years.

Special thanks from both of us go to the crew at Columbia University Press, including Science Publisher Patrick Fitzgerald, Editorial Assistant Ryan Groendyk, Senior Manuscript Editor Irene Pavitt, Senior Designer Milenda Lee, Publicity Director Meredith Howard, and Senior Publicist Peter Barrett, as well as to Annie Barva, for her meticulous copy editing.

Our deep appreciation goes to all who provided invaluable comments on various drafts of the book, including Aaron Huertas, Susan Joy-Hassol, Joseph Romm, Matthew Nesbitt, and Jonathan Cobb.

The Madhouse Effect

PEOPLE AREN'T LAUGHING AT YOU BECAUSE YOU'RE GALILEO.
THEY'RE LAUGHING BECAUSE YOU'RE A CRANK DRESSED UP PRETENDING TO BE GALILEO.
TOLES

[1]

SCIENCE

How It Works

Science. Everybody says they are for it. So why the firestorm of argument about the science of climate change? It's an interesting question. With a disturbing answer. For all the complexity of detail in science, the process is actually fairly straightforward.

Science is unique among human endeavors in the "self-correcting" machinery (to quote the famous Carl Sagan)[1] by which it is governed. That machinery ensures that science continues on a path toward an increasingly better understanding of the natural world despite the occasional wrong turns, dead ends, and missteps. The machinery consists of the critical checks that exist in the form of peer review and professional challenges, with the overriding maxim—again attributed to Sagan[2]—that *extraordinary claims* especially require *extraordinary evidence*. Good-faith skepticism—that is, skepticism that attempts to hold science to the highest possible standard through independent scrutiny and questioning of every minute detail—is not only a good thing in science but, in fact, essential. It is the lubricant that ensures that the self-correcting machinery continues to function.

Unfortunately, the term *skeptic* has been hijacked, especially in the climate change debate, to mean something entirely different. It is used as a way to dodge evidence that one simply doesn't like. That, however, is not skepticism but rather contrarianism or denialism, the wholesale rejection of validated, widely accepted scientific principles on the basis of opinion, ideology, financial interest, self-interest, or all these things together.

We must distinguish true skepticism—a noble attribute found in all good science and all good scientists—from the pseudoskepticism practiced by armchair science critics who misguidedly fancy themselves modern-day Galileos. As Carl Sagan also once said, "The fact that some geniuses were laughed at does not imply that all who are laughed at are geniuses. They laughed at Columbus, they laughed at Fulton, they laughed at the Wright Brothers. But they also laughed at Bozo the Clown."[3] For every Galileo, there are many thousands of Bozo the Clowns. When it comes to the fractious debate over policy-relevant areas of science, the Bozos are too often the ones with the megaphones.

Skepticism

True scientific skepticism takes many forms. Skepticism takes place in the give-and-take at scientific meetings, where scientists present their findings and then address questions, criticisms, and challenges from their colleagues in the audience. It takes place in the form of peer review: Scientists write up their findings and submit them to journals. The journals select several other scientists with expertise in the field to critically evaluate the submission. If they find flaws in the data, the underlying assumptions, the experimental design, or the logic, the authors must revise and resubmit. This process might be repeated a number of times for a single article. In the end, the article is published if and only if the editor determines that the authors have satisfactorily addressed any concerns or critiques raised during the review process and that the manuscript represents a positive contribution to the existing scientific literature.

The quality-control process of peer review isn't perfect, of course, and flawed work inevitably does get published. Certainly, no single scientific article ever defines the collective body of knowledge. And so there is even peer review of peer review in the form of multiauthored scientific

assessments, like those by the National Academy of Sciences, that evaluate the collective evidence in the peer-reviewed literature on a particular topic and summarize the state of knowledge on the topic. These assessments, too, are peer reviewed for accuracy, objectivity, and thoroughness.

The fact of the matter, however, is that there is a weakness in the scientific system that can be exploited. The weakness is in the public understanding of science, which turns out to be crucial for translating science into public policy. Deliberate confusion can be sown under a false pretext of "skepticism." And the scientific process is continually under assault by bad-faith doubt mongers.

There is, for example, the whole duplicitous game of assigning motives. Critics sometimes seek to cast suspicion on the scientific enterprise by suggesting that it is compromised by a conspiracy of ulterior-motive-driven individuals. "The scientists are in it for the money," they say, seeking to "get rich off of government grant money." Ironically, there is no small amount of *projection* here. These accusations, after all, are typically made by talking heads who get paid hefty sums by industry front groups to peddle disinformation to the public and to attack the scientists.[4] But what about the substance of the accusation? Do climate scientists, for example, seek to reinforce the dominant narrative that climate change is real and caused by humans to generate concern from the public and policy makers just so they can guarantee the ongoing availability of government grant support for their work?

To understand just how absurd that premise is, we have to understand something about how science really works. In science, you don't make a name for yourself by simply reinforcing the dominant narrative. You don't get articles published in the premier journals *Nature* and *Science* by simply showing that others were right. The way you establish a name for yourself in the world of science is by demonstrating something new or surprising, by contradicting conventional wisdom. A record of novel, groundbreaking work is what gets you tenure, what helps bring in research grants, what leads to salary increases from your institution.

Any scientist who could soundly demonstrate that Earth is *not* warming would become an instant science celebrity. A scientist who could definitively explain the warming of Earth by natural rather than human causes would have prominent articles published in *Nature* and *Science*. He or she would appear on the network news and make the cover of *Scientific American*. Such an individual would be tenured, promoted, likely elected to the National Academy of Sciences. The scientist would go down in history as one of the great paradigm breakers of all time, part of the exclusive club that includes Galileo, Newton, Darwin, Einstein, and Wegener (of plate tectonics fame). Such a scientist would, in short, achieve both fame and fortune.

So the incentives for a scientist would appear to be rather the opposite of what the critics claim. But let us not forget that in science the more extraordinary the claim, the more extraordinary the evidence must be. If you are a scientist who publishes a bold new claim, you better be prepared to defend it scientifically, for there will be a proverbial target on your back. Other scientists will be gunning for you. Discrediting your famous new finding might be just the ticket to their own fame and glory. Remember "cold fusion"—the controversial claim by a pair of chemists in the late 1980s that it was possible to generate fusion energy at room temperature using little more than tap water and a couple of electrodes? The American Physical Society devoted a whole session at its annual meeting to debunking it. A group of Caltech physicists were celebrated by the scientific community and gained quite a bit of public exposure for discrediting the claim. But they, in turn, had to prove their case. They had to present compelling evidence that the "cold fusion" claim was wrong, which they did.[5] That's the way it works.

The Hockey Stick

We would be remiss in this context not to recount the tale of the hockey stick. In the late 1990s, one of the two authors of this book (Michael Mann) published the now-famous "hockey stick curve" depicting temperature trends over the past millennium. The curve demonstrated the unprecedented nature of modern global warming. It became a symbol in the climate change debate and thus a potent object of attack to some.[6] Given the iconic status of the hockey stick curve, there was a huge potential reward in terms of scientific notoriety for any scientist who could demonstrate it wrong. Many have indeed tried.

Dozens of scientific groups have performed their own studies, using different data and different methods and coming to their own independent conclusions. Challenges to the hockey stick curve have been published in the leading scientific journals, such as *Nature* and *Science*. Some of these challenges helped launch the careers of ambitious young scientists. Yet a decade and a half later, a veritable hockey league of studies confirms the basic hockey stick conclusion.[7] The most comprehensive study to date has yielded a curve that is virtually indistinguishable from the original hockey stick curve.[8] The basic finding has stood the test of time. Scientists have now largely moved on, seeking answers to other related questions (for example, What natural factors primarily drove prehistoric temperature changes?). That's how science works. What once lay at the scientific forefront—the speculative boundary of scientific understanding—is slowly absorbed into the corpus of scientific knowledge. But if—and only if—it stands up to repeated challenge. The boundary of scientific knowledge expands, and science moves on, exploring new frontiers. If the original hockey stick curve had indeed been wrong, we would know by now. If assertions of global warming were wrong, we would know by now. If global warming were *not* caused by humans, we would know by now. The incentive structure in science is such that establishing any of those propositions—were it possible—would be irresistible to some aspiring young scientist looking to establish his or her name.

That is not at all to say that efforts to discredit the hockey stick curve or assertions of global warming or the human cause of global warming have ceased. Quite the contrary. Although the scientific community may consider these points to be settled fact and may have moved on, there are others who have not. Powerful vested interests that consider such findings rather inconvenient fight on. Just as the tobacco industry found medical research revealing the damaging health impacts of cigarette smoking to be a threat to its interests and so recruited contrarian scientists, think tanks, and lobbying firms to participate in a massive public disinformation campaign aimed at undermining the public's faith in the scientific evidence,[9] and just as the chemical industry has continued to seek to discredit science demonstrating negative health and environmental impacts of their products, fossil fuel interests have continued to spend tens of millions of dollars in an ongoing public-relations campaign to discredit the science of human-caused climate change, including the hockey stick curve itself.[10] These attacks are part of what has been called a "war on science,"[11] a war waged by special interests seeking to avoid oversight and regulation in the face of compelling scientific evidence that exposes the risks created by their products, actions, and services.

The Bizarro World

The late senator Daniel P. Moynihan (D-N.Y.) once famously quipped, "You're entitled to your own opinions but not your own facts." Unfortunately, in today's world many *do* feel entitled to their own facts—particularly when those alternative "facts" support an agenda. If you read the *Wall Street Journal* editorial page or tune into Fox News, you find yourself in a Bizarro world where up is down, left is right, and black is white. Ozone depletion is a myth. Pollution? Shmolution! And evolution? Just a *theory*. Acid rain is caused by trees. And global warming? Don't even get us *started* on global warming!

A growing number of American citizens find themselves trapped within the Bizarro world's bubble of misinformation. Gone are the days of Walter Cronkite when "that's the way it was," when we could agree on the facts and politely agree to disagree on their implications. Nowadays, many of us choose to access only those sources of information (networks, websites, newspapers, radio shows) that simply reinforce our preconceptions. If you're a fervent climate change denier, there is a good chance that you (1) aren't reading this book anyway, (2) don't read Tom's editorial cartoons in the *Washington Post* or Mike's commentaries and interviews, and (3) get your information about climate change from conservative media outlets committed to perpetuating the notion that climate change is a myth, a vast conspiracy by thousands of scientists around the world perpetrating a massive hoax to create a new socialist world order.

If that is your viewpoint, it probably doesn't matter to you that the world's scientists have reached an overwhelming consensus that climate change is (1) real, (2) caused by us, and (3) already a problem. It probably doesn't matter that the National Academy of Sciences, founded by (Republican) President Abraham Lincoln, has stated this to be the case, as have the scientific academies of all the major industrial nations.[12] It probably doesn't matter that every scientific society in the United States to weigh in on the matter has done so as well.[13] From your perspective, that's all just evidence that the scientific conspiracy must run even wider

and deeper. It is this sort of epistemic closure that makes it increasingly difficult to reach hardened climate change deniers.

Science is litigated through the formulation and testing of hypotheses, the analysis of hard data, and an examination of the facts, not by television debates between talking heads with opposing viewpoints. Yet our mass media too often frame climate change precisely that way. Who do you imagine benefits from that framing?

Doubt Is Their Product

When it comes to the public battle over policy-relevant science, special interests have long recognized that they enjoy the advantage of the prosecution in the court of public opinion. Their internal research, focus groups, and polling have revealed that they need generate only sufficient uncertainty about the scientific evidence in the public mind-set to ensure an agenda of inaction. "Doubt is our product" read one internal tobacco industry memo.[14]

It is unsurprising that right-wing media outlets serve as mouthpieces for this agenda. But what is more troubling is that the *mainstream* media have often played the role of unwitting accomplices. By perpetuating the notion that there are two equal "sides" when it comes to objective matters of fact such as evolution and climate change, mainstream-media outlets have reinforced the perception that there *is* legitimate doubt about these matters. Comedian John Oliver had it precisely right when he brought ninety-seven scientists out onto the studio floor (including our friend Bill Nye, "the Science Guy") to debate three climate change deniers.[15] The split is ninety-seven–three, not fifty–fifty, when it comes to support among actual publishing scientists for the scientific consensus on human-caused climate change.[16]

As we shall see, the success of the industry-funded climate change–denial machine derives in part from media outlets' willingness to emphasize conflict over consensus, controversy over comprehension. Add in a bit of Journalism 101 false "balance," and you have a perfect recipe for industry front groups and their hired guns to bake up mass confusion and to do what they do best—obfuscate and obscure the facts and delay action.

The increasingly popular refrain among politicians, "I am not a scientist," is just another formulation used to avoid an intelligent discussion of climate. It is not in any way a coherent response because logically the follow-up would be "so I will defer to the consensus of people who *are* scientists." But the politicians never seem to get to that part.

They instead act as though unless there is 100 percent unanimity among scientists, nothing can be known. But they are also contradictorily fully willing to assent to a tiny minority of scientists who happen to be saying what they want to hear. As often as not, the scientists in question are not even climate scientists, but who cares? Parts is parts.

For example, if you carefully select a specific group of years of temperature readings from the past decade or two, you could try to argue that there was a "pause" in warming. The cherry-picking of the starting year in the sequence betrays the lack of sincerity of those advancing this argument, which has come to be known as the "faux pause" in scientific circles.[17] But all a politician who denies climate change needs is one scientist to go on record with the "global warming has stopped" talking point. The politician is then perfectly willing to cite that claim as conclusive evidence, trumpet the "fact" that warming had ceased altogether, and proclaim that there's nothing to worry about.

Of course, further data have shown that any slowdown in warming was fleeting. There was no "pause," and heat records continue to be shattered. Confronted with this fact, denialist politicians merely shift the goalposts: "Oh well, climate changes all the time."

Avoiding the Abyss

So how is the public supposed to navigate this morass of fact and opinion? There will always be uncertainty, but it is hardly a reason for inaction. To believe otherwise is to buy into the fallacy that because we don't know everything, we know nothing. We make decisions in the face of uncertainty in every aspect of modern life. We don't refuse to fly on an airplane, drive a car, cross a street, breath the microbe-infused air simply because there is a very slight risk of harm in each of these actions.

When it comes to global environmental threats—such as acid rain, ozone depletion, and climate change—both the probability and the consequence of harm are far greater. An understanding of the basic facts and a rudimentary respect for the principle of precaution are more than adequate causes for action.

Two basic facts underlay climate science, and they were there from the beginning. First, carbon dioxide (CO_2) was well known to be a heat-trapping gas. Second, humans were significantly increasing the amount of CO_2 in Earth's atmosphere through the burning of fossil fuels and other activities. We were in fact well on our way to *doubling* it. Right there, we

had a prima facie case that we were warming the planet and affecting Earth's climate system.

But the public conversation was soon diverted into a discussion of various hypotheticals. Maybe there would be a negative-feedback loop in which more trapped heat would lead to increased cloud formation, and therefore more of the Earth-warming sunlight would bounce back into space. But for every new possible negative-feedback-loop scenario, there was a potential positive one, too: a warming permafrost, for example, would release trapped methane, yet *another* heat trapper. In fact, the positive feedbacks won out.

In the face of overwhelming evidence that (1) CO_2 traps heat, (2) we're on track to double CO_2 levels by midcentury, and (3) unprecedented change in our climate will likely result if we remain on that trajectory, the burden of proof really ought to have fallen on the side of those asserting the contrary. That is what we call the *precautionary principle*. The burden was on the critics to prove that, given these known facts, doubling the amount of CO_2 in the atmosphere in less than two centuries (a mere instant in geological time) would have no meaningful impacts. Instead, we got a decades-long stream of increasingly far-fetched hypotheticals intentionally designed to obscure a rather straightforward set of facts.

So what to do? What to think? Two things at minimum: Where there is doubt about the facts, if you do have some respect for the scientific method, look to the scientists who are doing the hard work of measuring, analyzing, and understanding. Where there is apparent disagreement or continuing uncertainty, go with the preponderance of evidence. It's not all that hard. Repeat after us: "preponderance of evidence."

Absolute proof is for mathematical theorems and alcoholic beverages. Those who demand it when it comes to matters of science are displaying a disregard for the scientific process. Science deals instead with degrees of likelihood, balances of evidence, and consistency among lines of evidence. We can't "prove" gravity. It is, after all, *only* a "theory." But when hiking a knife-edge mountain trail, we respect it nonetheless. We can't "prove" evolution. It, too, is a *just* a "theory." But without our understanding of that theory, we would be losing our battle against infectious diseases such as the flu.

Nor can we "prove" the greenhouse effect. It, once again, is a *mere* "theory." That didn't stop the U.S. Air Force from using its understanding of the greenhouse effect and atmospheric absorption of heat when it developed heat-seeking missiles in the 1950s.

So we should have our fullest respect for the scientific framework behind the proposition that the burning of fossil fuels and other activities are changing Earth's climate. The evidence is overwhelming, and it has only increased in strength and consistency over time—the hallmark of a compelling scientific framework.

To formulate a policy response to climate change, we didn't have to be 100 percent certain. There were numerous policies we could have enacted to slow the increase of CO_2 in the atmosphere, and they would have been inexpensive compared with the cost of climate change damage. Fuel- and energy-efficiency options were readily available and would have had independent benefits. The costs of inaction were already far greater than the cost of taking action, and over the years they have become only greater.

Decades ago, time was still on our side. Reductions then would have meant fewer reductions needed now.

Well, we ignored the science, and we avoided the sensible choices that were before us. And now we are already paying the price. Time is no longer on our side. Let's use what time we have more wisely.

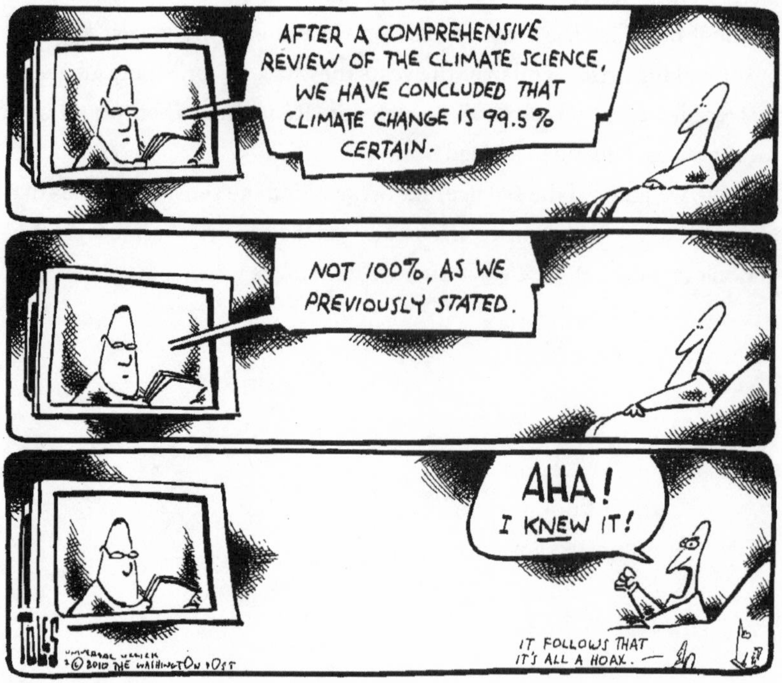
AFTER A COMPREHENSIVE REVIEW OF THE CLIMATE SCIENCE, WE HAVE CONCLUDED THAT CLIMATE CHANGE IS 99.5% CERTAIN.
NOT 100%, AS WE PREVIOUSLY STATED.
AHA! I KNEW IT!
IT FOLLOWS THAT IT'S ALL A HOAX.
TOLES
© 2010 THE WASHINGTON POST

2-22-10

[2]

CLIMATE CHANGE

The Basics

The basics of climate science are actually very simple and always have been. Carbon dioxide in the atmosphere traps heat, and we are adding more CO_2 to the atmosphere. The rest is details.

It is important to separate the essential part of a scientific argument from the often exceedingly complex details. For example, everyone understands now that Earth is a sphere and revolves around the sun. That's the simple part. Describing the actual shape of Earth (an "oblate spheroid") and its orbit (an "eccentric ellipse") is more complicated. And making calculations about orbits of planets, timing of eclipses, or trajectories of spaceflights is complicated and difficult to do. But you don't have to do the calculations yourself to understand either the basic ideas at work or the results of the calculations.

So, too, it is with climate. The climate change–denial industry has adopted the strategy of obscuring the basic concepts through a torrent of typically misleading arguments about technical details and minutia. But all of that never has and never will change the basic fact that more CO_2 in the atmosphere traps more heat and warms Earth's surface.

There is a difference between simple and simplistic. Here's simplistic: Exclaiming "Where's your global warming NOW?" in response to a cold day in winter. Or reacting to a *warm* day in winter with a derisive "If this is global warming, I'LL TAKE IT!" Or bringing a snowball onto the floor of the Senate as ostensible disproof of global warming. Or conceding that measurements do in fact show that Earth is getting hotter by proclaiming, "Climate changes all the time!" Yes, in fact it *is* and *has been* changing all the time, and we'll get into that later. But *this time* we're the ones who are creating the change, and this change is for the worse, not the better.

Physics and Chemistry 101

Next time that cantankerous uncle of yours whom you see every Thanksgiving tells you that the greenhouse effect is "controversial new science," remind him that it's actually basic physics and chemistry that go back nearly two centuries. Joseph Fourier—the same scientist who discovered the *law of heat conduction* and the Fourier Series of mathematics—understood that certain gases in the atmosphere, such as CO_2, have a heat-trapping property that we have come to know as the "greenhouse effect."

Svante Arrhenius—the same scientist who more than a century ago gave us the *definition of an acid*—already recognized that we were increasing the amount of CO_2 in the atmosphere through the burning of fossil fuels and that Earth's temperature was rising in response. And since we mentioned the topic of acids, let's take note that the same CO_2 buildup is also acidifying our oceans (that "other CO_2 problem") through basic chemical processes that Arrhenius well understood.

Observations of modern air samples that have been collected for more than half a century and samples of ice cores that are many thousands of years old tell us that we have now reached CO_2 levels in the atmosphere not seen since the dawn of human civilization and perhaps even in 5 million years.

Before the Industrial Revolution, there were about 280 parts CO_2 per million parts of atmosphere (ppm). We have now passed the 400 ppm mark. Earth has warmed around 1.5°F (1°C) in response. That warming is just the veritable tip of the iceberg.

At the current rate that we are burning fossil fuels, we will have doubled CO_2 concentrations (that is, approached around 550 ppm) by midcentury. Arrhenius estimated that this doubling would raise Earth's temperature

around 9°F $(5°C)$.[1] We will forgive him if his nineteenth-century paper-and-pencil calculations weren't as precise as those based on climate model simulations using the most powerful supercomputers in the world, which give a modestly lower estimate of around 5.5°F $(3°C)$ (this quantity is known as the *equilibrium climate sensitivity*, a measure of how sensitive Earth's climate is to increases in greenhouse gas concentrations). That is still roughly *double* the amount of warming we would expect from the greenhouse effect of the added CO_2 alone and is a result of the *positive-feedback loops* discussed in chapter 1, which have the net effect of amplifying the warming.

The same models tell us that (1) we can't explain the observed warming from natural causes—only the human influence of increasing greenhouse gases matches the pattern of warming in the observations—and (2) under business-as-usual fossil fuel burning, we will have indeed raised Earth's temperature by around 9°F $(5°C)$ by the end of the twenty-first century.

That's roughly the same amount of warming that took place from the height of the most recent ice age (around 25,000 years ago)—when an ice sheet covered what is now Manhattan—to today.

If you're with us so far, then you understand the basic physical science behind global warming.[2] That wasn't so hard, was it?

Hotter, Drier, and . . . Wetter?

Now, what else do we expect as a result of this warming? Naturally, we expect to see more frequent and more devastating heat waves like those that have already afflicted the United States, Europe, India, Pakistan, and other regions across the globe in recent years. But extreme heat is just the most obvious symptom of our planet's fever.

As the climate warms in response to increasing carbon emissions, large-scale wind patterns are expected to shift. The region of sinking dry air currently found in the subtropics (think Sahara and Mojave Deserts) will continue to shift poleward, meaning drier, more desert-like summer conditions in regions such as the United States. Although it might seem a paradox, we also expect worse flooding in many of the same regions. "How's that?" you ask.

Warmer air holds more moisture than cooler air; in fact, the increase is roughly exponential—a consequence, once again, of very basic physics and chemistry. So we have moisture in the air that can cool and condense and produce rain or snow when there is *rising* motion in the atmosphere. That last bit is crucial. In regions where the tendency is instead toward *sinking* motion, those conditions may become less frequent, especially in summer. But when they *do* occur, there is the potential for far greater amounts of precipitation. It may rain less often, but *when* it rains, it will literally pour. That is why it is possible for Texas to have experienced the twin plagues of record drought (in the summer of 2011) and record flooding (in the spring of 2015) in recent years.

Then there is the epic, seemingly unending drought in California, now going on its fifth year.

Despite some debate within the scientific community about the extent to which climate change is responsible for the unusual jet-stream pattern steering the moisture-bearing storms northward, it is the combination of unusually low precipitation and record temperatures, causing what little moisture soils do contain to evaporate, that is responsible for what appears to be the worst drought in California in more than a millennium. That combination appears to bear the fingerprint of human-caused climate change.[3]

Rising Oceans

As glaciers and ice sheets melt the world over, and as seawater literally expands as warming penetrates down into depths of the ocean, global sea levels will continue to rise. The roughly *10 inches* (25.4 centimeters) of global sea-level rise we have seen thus far is, once again, the tip of the proverbial iceberg. As the Greenland and West Antarctic Ice Sheets begin to melt—something that was predicted to take many decades but is now already being observed—sea levels will rise faster and faster. The acceleration is under way.

Even the more conservative estimates now have global sea level rising about 3 feet (1 meter) by the end of this century, with *5 to 6 feet* (1.5–1.8 meters) difficult to rule out. Given that less than 1 foot (30 centimeters) of sea-level rise is already posing a severe threat to low-lying island nations such as Tuvalu, Kiribati, and the Maldives in the tropical Pacific and Kivalina in the Arctic as well as to coastal regions around the world from Bangladesh to Miami Beach, one can only imagine what 5 to 6 feet would do.

Contrast the Arctic with the Antarctic. In the Arctic, we have an ocean rather than a continent, with surface ice (sea ice) that expands in winter and contracts in summer. Loss of that ice will not contribute in any meaningful way to sea-level rise, much as a melting ice cube floating in a glass of water doesn't raise the water level as it melts. But it does pose problems of its own.

A delicate food web relies on the Arctic sea ice environment, from the microorganisms that live in the Arctic Ocean all the way up to the charismatic megafauna we know as the walrus and the polar bear. Get rid of the sea ice, and you disrupt that food web. More on that later.

Climate models have predicted that Arctic sea ice will be absent during summer by the end of this century. Actual observations, however, point to a more alarming time line. With the precipitous decline seen over the past decade, many experts predict that we'll see ice-free summers in the Arctic in a couple of decades.

That decline already appears to be having an impact on regions far afield from the Arctic itself, wreaking havoc from California to New England. Studies going back more than a decade show that reduced Arctic sea ice may be favoring the increasingly persistent and rather unusual jet-stream pattern that is exacerbating the ongoing drought in California in the form of a "ridge" of high pressure that nudges the rain- or snow-producing storms northward.[4] The same jet-stream pattern led to temperature extremes (of warm and cold) across the United States in recent winters and may very well have favored the fateful anomalous westward track that led Superstorm Sandy directly toward New York City and the New Jersey coast, when most such storms head out to sea.[5]

The Day After Tomorrow—a Possibility?

Climate change may also be disrupting key ocean currents. You may recall the movie *The Day After Tomorrow* (Roland Emmerich, 2004). Well, rest assured that tornadoes will not wreck Los Angeles, giant megahurricanes will not cover the entire Northern Hemisphere, and continental ice sheets take thousands of years, not days, to form. Those details notwithstanding, there is a small grain of truth to the premise underlying the movie.

Ice ages come and go on very long timescales—over tens to hundreds of thousands of years—driven by modest adjustments in Earth's orbit around the sun that matter little over years or centuries but add up to significant changes over many millennia. Over the past 700,000 or so years, the ice ages have come and gone roughly every 100,000 years. The pattern consists of brief warm *interglacial* intervals (such as the so-called Holocene period, which has prevailed over the past 11,500 years or so) of relatively moderate global temperatures and greatly diminished continental ice sheets alternating with relatively long *glacial* intervals of extensive ice cover and relatively cold global temperatures (around 7–9°F [4–5°C] cooler than today). The buildup of ice as we head toward an ice age takes place very slowly, over tens of thousands of years, but the terminations can be extremely rapid, taking only a few thousand years.

The most recent ice age began a little more than 100,000 years ago, and the termination began around 13,000 years ago. But a funny thing happened on the way to the Holocene. Much of the water produced from the melting ice flowed into the subpolar regions of the North Atlantic Ocean. Once there, it freshened the typically salty and thus dense seawater that is typically formed there. Freshwater is lighter than salty water, so this infusion of freshwater inhibited the sinking of water that typically takes place in the subpolar North Atlantic. Here's where things start to get interesting.

That sinking motion actually helps drive a very large-scale and important current system that we call the North Atlantic Drift. In common parlance, it is referred to as the Gulf Stream, but oceanographic purists reserve that name for the wind-driven current system found off the southeastern coast of the United States, which leaves the coastline around Cape Hatteras, North Carolina, and heads off toward Iceland and Europe, where it indeed *becomes* the North Atlantic Drift. The North Atlantic Drift, in reality, is best thought of as the surface current in a conveyor-belt-like pattern of ocean circulation in which warm surface waters are carried north into the higher latitudes of the North Atlantic, helping to keep that region warmer than it would otherwise be. Those surface waters then sink in the subpolar North Atlantic, head south in the deep ocean, and eventually surface again at lower latitudes, eventually completing the circuit by returning to the North Atlantic.

Shut off that conveyer-belt ocean-circulation pattern, and you cool the North Atlantic and some neighboring regions. That's what happened when all the meltwater arrived in the North Atlantic 13,000 years ago. The North Atlantic and neighboring regions in North America and Europe appear to have been nudged back into ice-age-like conditions for about 1,500 years before the final exit from the ice age 11,000 years ago. That approximately 1,500-year interval is referred to as the Younger Dryas because of the species of tundra flower (*Dryas*) found in the remains of European pollen dating to that period.

If the meltwater from a glacial termination was enough to send at least part of the Northern Hemisphere back into ice-age-like conditions at the end of the most recent ice age, perhaps the melting of Northern Hemisphere glaciers (especially the Greenland Ice Sheet) from global warming may do something similar? That is indeed the premise that underlies *The Day After Tomorrow*. In reality, there is too little ice now to yield anything nearly as dramatic as the Younger Dryas. We're not in store for another ice age. But climate model simulations suggest the possibility of a substantial weakening of the conveyor belt by the end of the century. And, once again, the observations seem to suggest that we're further along than the models predict we should be, with evidence of a substantial weakening already apparent.[6]

What we'll likely see at most is the cooling of a small patch of ocean south of Greenland and a bit less warming than we would otherwise expect in some maritime regions neighboring the North Atlantic. However, the disruption of this ocean-current system could adversely affect marine productivity in the North Atlantic, a major source of global fish stocks.

So if we won't be devastated by massive tornados in Los Angeles or be plagued by continent-size hurricanes, might we not expect at least *some* trend toward more extreme weather?

In some cases, the answer is simple. As mentioned earlier, we expect more frequent, more intense heat waves in a warmer world. But cold extremes on balance should diminish. With a warmer, more moisture-laden atmosphere, storms will produce more precipitation. More spring and summer flooding, more of those paralyzing coastal blizzards during the increasingly short winter. Severe summer thunderstorms? A warmer, more humid atmosphere should yield more of them. And with them, more destructive wind gusts, downpours, and hail.

So that means worse tornadoes, too, right? Not so fast. Scientists are still debating that question because it turns out to be rather complicated. Tornadoes feed on what is known as "convective available potential energy," or CAPE (that's nerd talk for warmer, moister, less-stable atmosphere conditions—the type of conditions that are favorable for thunderstorms). Another factor is required to produce a tornado: twisting winds, which rely on something known as *wind shear* (when the wind speed or direction changes with height). The so-called Tornado Alley (a broad north–south swath cutting through Texas, Oklahoma, Kansas, Nebraska, and South Dakota) is a region where these factors often coincide during late spring.

It is possible that climate change could cause CAPE to increase but wind shear to decrease in tornado-prone regions. In that scenario, we might see little or no change in tornado activity. Some experts are finding, however, that increased CAPE is likely to dominate over any changes in wind shear and that we'll likely see an uptick in the intensity of tornadoes—that is, more of the F4 and F5 monsters (the strongest two categories in the Fujita scale) like the one that killed dozens of people in Oklahoma City in May 2013.[7] But this possible outcome is at the more speculative boundary of our current scientific understanding.

Tempests in a Greenhouse

And while we're on the subject of deadly storms, what about the deadliest storms of all—hurricanes? Here, too, there is a debate within the scientific community because multiple factors are at work that could in principle have offsetting effects. On the one hand, warmer oceans, on balance, are likely to produce more intense hurricanes and supertyphoons because there is more energy—in the form of warm, moist air—available to intensify these storms. That means more Katrinas and more Haiyans.

On the other hand, changes in wind shear once again play a role. In this case, wind shear is actually a *bad* thing. It disturbs the vertical, cylinder-like structure of a storm because winds are pushing the storm one way near the surface and another way at a higher altitude. The El Niño phenomenon—the warming of ocean waters in the central and eastern equatorial Pacific that comes every few years, impacting climate around much of the rest of the world—affects the wind shear. There tend, for example, to be fewer tropical storms in the Atlantic during El Niño years when wind shear in the Caribbean and tropical Atlantic is higher than normal and more tropical storms during the opposite La Niña state.

If climate change increases the amount of wind shear (for example, in the tropical Atlantic), that increase may create a less-hospitable environment for tropical storms and hurricanes. One possible scenario is a more El Niño–like future (most climate models indeed predict this outcome, though it is plausible the models have that wrong),[8] more wind shear in the tropical Atlantic, and fewer Atlantic tropical storms. But those storms that do form, boosted by the warmer ocean, will tend to be stronger. Some leading researchers in the field have argued that *both* the number and the intensity of these storms will increase in a warmer world.[9]

So there is uncertainty here. But that doesn't mean we should fall for the "because we don't know everything, we know nothing" fallacy. Asking whether climate change "caused" a particular storm is a particular pet peeve of ours. It's the wrong question. It's like asking which of Barry Bond's record seventy-three home runs in 2001 were caused by steroid

use. It's the same loophole that the tobacco industry attempted to use to avoid accountability for a product that killed millions of people. And it's this loophole that polluting interests are using to escape culpability for the damage that burning fossil fuels is doing to our planet.

A Loophole You Could Lose a Planet Through

We of course can't say that climate change "caused" a particular heat wave, flood, or storm. There is always the chance that the heat wave, flood, or storm would have happened anyway. But climate change is almost certainly making these events more frequent. There is an increased occurrence of these events because of climate change, just as there is an increased incidence of lung cancer among smokers and an increased number of home runs by steroid-using baseball players.

Individual events, moreover, are being made demonstrably *more* extreme and *more* damaging than they otherwise would have been.[10] Let's consider Superstorm Sandy in 2012, for example. If nothing else, the around 1 foot (30 centimeters) of sea-level rise off the mid-Atlantic coast means that Battery Park in New York City saw a 14-foot (4.3 meter) rather than a 12-foot (3.6-meter) peak storm surge. That increment may seem small, but it was enough to yield 25 square miles (65 square kilometers) of additional flooding along the New York and New Jersey coasts, which affected the homes of more than 80,000 additional people[11] and cost more than $2 billion in additional damage.[12] That's just the damage due to sea-level rise. It doesn't even account for the fact that the record intensity (for a storm that far north in the Atlantic), unprecedented size, and unusual track of Sandy may also have been influenced by climate change.

Then there is Hurricane Irene. Irene was not caused by climate change. It was actually a rather run-of-the-mill, late-summer Atlantic hurricane. What was not run of the mill was the amount of rainfall it produced, yielding record flooding throughout a broad swath stretching from eastern Pennsylvania to northern New England. That flooding can be tied, at least in part, to record-level surface temperatures in the Atlantic Ocean, which meant that there was more moisture in the air for the storm to entrain as it spun up off the mid-Atlantic coast, moisture that was turned into record rainfall inland.

And the record 108 or more inches (274 centimeters) of snowfall that brought Boston to a standstill in the winter of 2014/2015? That record snowfall was not, as the critics would have you think, evidence *against* climate change but rather evidence *of* it. These huge snowfalls resulted from a series of very strong Nor'easters that intensified over unusually warm midwinter seas off the New England coast. Once again, that meant more moisture available for the atmosphere, in this case, to turn into extreme snowfalls.

Tipping Points: Is It Too Late?

When we look at the increasingly profound impacts that climate change is now having on us and our environment, the problem can seem pretty overwhelming. People understandably often ask, "Is it too late now? Have we passed the *tipping point*?" The answer is yes, no, and maybe.

A tipping point is, of course, a point of no return. In the context of climate change, it would mean that we have warmed the planet enough to set in motion an unstoppable process. In reality, there is no single tipping point in the climate system; there are many. And the farther we go down the fossil fuel highway, the more tipping points we will cross. Many observers have argued that a warming of the planet of 3.6°F (2°C) relative to preindustrial levels (something that will likely happen if we allow CO_2 levels to climb to just 450 ppm) would almost certainly create dangerous, potentially irreversible changes in our climate. As a reminder, we have already warmed around 1.5°F (1°C), and another 0.9°F (0.5°C) is likely in the pipeline. Another decade of business-as-usual fossil fuel emissions could commit us to that 3.6°F (2°C) "dangerous warming" threshold.[13]

We might have already crossed at least one key tipping point: most, if not all, of the West Antarctic Ice Sheet now appears on course to disintegrate no matter what we do—the warming of the Southern Ocean has eroded the stabilizing ice shelves to the point where there may be nothing to hold the land ice from calving inexorably into the ocean in the decades ahead. That would mean another 10 feet (3 meters) or more of sea-level rise on top of what we are already in store for.[14]

This process could take a millennium to unfold, but we can't rule out the possibility that it may happen faster than that. Maybe two centuries. Maybe one. There is still a fair amount of uncertainty here. But we have already seen that uncertainty is not our friend. Many impacts, such as the melting of the ice sheets and the loss of Arctic sea ice, appear to be happening *more quickly* than we expected. Uncertainty seems to be breaking against us, not for us, in many respects.

There are, however, other tipping points we may still be able to avoid. We have likely not yet committed to the melting of the majority of the Greenland Ice Sheet, which, in melting, would give us at least another 10 feet (3 meters) of sea-level rise. Nor have we likely committed to the melting of the East Antarctic Ice Sheet, which would give us a massive 200 feet (60 meters) of sea-level rise. Some experts, however, have argued that even 3.6°F (2°C) of planetary warming could indeed guarantee that substantial melting of this ice sheet, too, will take place.[15]

In all likelihood, many other tipping points lie out there like land mines, but we don't know exactly where they lie or precisely how much additional warming will trigger them. We are the blindfolded man who is told he is nearing the edge of a cliff. Is he three steps away? Four? Ten? Regardless of the distance, his only safe course of action is to stop lurching forward.

Can we stop our forward lurch before it is too late? Perhaps, but only if we experience a *different* tipping point, a tipping point in the public consciousness; only if we collectively grasp the profligate nature of our continued reliance on fossil fuels and the urgency of transitioning rapidly toward a clean-energy economy; and only if we collectively get a satisfactory answer to the following question: Why should we give damn?

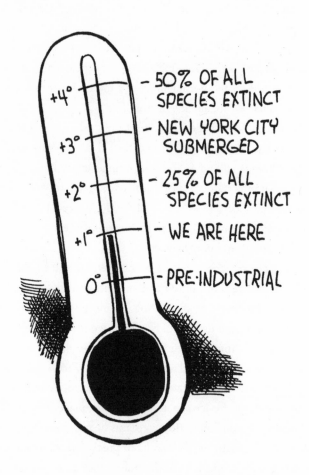

The Celsius temperature scale.

[3]

WHY SHOULD I GIVE A DAMN?

So why should we care? As it turns out, for pretty nearly every reason. The most unfortunate part of the societal debate over climate change has been the ease of imagining a warming climate as an essentially unthreatening occurrence, the status quo.

"Global warming" sounds almost pleasant. Like a day in springtime. You may have heard someone say in response to a nice day in winter, "If this is global warming, I'll take it!" A smooth, comfortable adjustment in living circumstances, not unlike turning up the thermostat by a degree or two. Not to worry!

This response is entirely understandable. Unfortunately, it is also entirely wrong.

Although the increase in CO_2 in the atmosphere is gradual and steady, the results of that increase will be anything but. People assume that noticeable effects will be far off in the future, but they are showing up right now. Relentlessly. And if you think the effects will be felt only in some faraway corner of the globe where only polar bears and penguins live, think again.

The consequences of a changing climate are occurring everywhere and, yes, likely right near you, affecting you, your family, your friends, your community. Be it national security, food, water, land, the economy, or health—ours and our planet's—the specter of climate change is upon us.[1]

Dreams of slowly adapting to climate change will have to be replaced with the hard reality of an ever-escalating pace of disruption and unpredictability.

In what ways will the effects of climate change be felt? In nearly every way.

Security

You don't have to be driven by ethics, morality, religion, or altruism to recognize the threat posed by climate change. Being a national security hawk, for instance, is more than adequate. In the lexicon of the national security community, climate change is the ultimate *threat multiplier*—it takes existing tensions and conflicts and amplifies them. A thawing Arctic Ocean means a new coastline to defend. It means open competition among the bordering nations of North America, Europe, and Asia for oil-drilling rights in the Arctic—rather ironic, since oil drilling is part of what got us into this mess in the first place. Climate change will create more competition among a growing global population for less food, less water, and less land—a prescription for a perfect storm of global conflict.

That storm is already brewing. A compelling case can be made that the historic ongoing drought in Syria was made worse by the aggravating effects of climate change and played a key role in the civil unrest and societal instability that ultimately led to the civil war there,[2] the reverberations of which continue to be felt around the world.

To understand how increased conflict might arise throughout the world, let's consider how climate change will affect the underlying contributing factors. Let's examine how climate change will have an impact on *all* sectors of our lives, from food, water, and land to human health, the stability of our fragile ecosystems, and, yes, our economy, too.

Food

The global population is currently around 7.3 billion and growing. It is projected to reach 9 billion by the middle of this century and could grow to 11 billion by the end of the century.[3] Malnutrition and hunger currently afflict more than 800,000 people, according to the World Food Programme.[4] That number will only get larger without a concomitant increase in food production. Yet climate change will likely lead to a *decrease* in global food production.

In the tropics, temperatures are already close to optimal for growing cereal crops such as rice, corn, and sorghum. That might sound good, but it's actually very bad. It means that even modest warming will lead to precipitous drops in yields due to the rapid descent down the far side of the productivity peak. That descent translates to a substantial decline in the cereal crops available to feed a large, growing, and too often malnourished population.

In the extratropics, the situation may seem better. Growing seasons will potentially get longer, and crops will potentially be grown at higher latitudes. That's the good news. Now the bad news. Any resulting gains could be more than offset by increases in damaging weather, more widespread and destructive wildfires, and longer and more frequent drought.

Think about the summer of 2012, when our nation's breadbasket was decimated by record heat and drought. Or think about the past five years, during which the unprecedented ongoing drought in California has threatened 33 percent of our total supply of fresh produce. In 2010, climate-related factors led to a 33 percent drop in wheat production in Russia, a 19 percent drop in Ukraine, a 14 percent drop in Canada, and a 9 percent drop in Australia.[5]

Less grain means less feedstock for livestock, and extreme summer temperatures mean greater heat stress on livestock, less water for them to drink, and fewer days and hours that farmers and ranchers can work the land. Ranchers in Oklahoma and Texas lost nearly 25 percent of their cattle during the record drought in 2011.

Well, at least we'll still have plentiful seafood, right? Alas, no. The disruption of ocean currents such as the North Atlantic Drift could impact the productivity of the North Atlantic, a critical source of fish and seafood. Meanwhile, we have seen massive die-offs of salmon in the Pacific

Northwest in recent years due to scorching-hot waters,[6] and West Coast oyster farms are getting hammered by global warming's twin brother: ocean acidification.[7] Rising CO_2 levels are likely to further exacerbate the threats to fish and shellfish stocks already threatened by overfishing, water pollution, and other human-caused threats.

Water

We human beings need food. But we can go without it for weeks. Water, however, we can go without for only a few days. And climate change means that there will be less of it—less freely available freshwater over a large part of the globe—to quench the collective thirst of a growing global population.

As the jet streams migrate poleward in a warmer planet, the dry sub-tropics will expand into the middle latitudes, leading to a wider zone of semiarid conditions. Now, that shift may seem to be balanced by other changes: the subpolar regions will likely receive more rain and snow, and the deep tropics, lying within the band of rising air currents known as the Intertropical Convergence Zone (ITCZ to weather nerds), may get even wetter as a warmer atmosphere produces greater amounts of rainfall. It sounds like a zero-sum game. Isn't the challenge just to move the water from where there is a surplus to where there is a deficit? Well, if that were all, then, yes, but the task would be extremely difficult and expensive, requiring massive investments in infrastructure. An ounce of prevention—in this case, cutting greenhouse gas emissions—is worth a pound of cure.

But there is a further complication. Just because more water falls doesn't mean that there is more moisture in the ground. Warmer soils and vegetation evaporate more moisture from the ground into the atmosphere. The current drought in California coincides with the warmest year (2014) on record there. That isn't a coincidence.

For the semiarid American West, the diminishing water supply and the growing population are on a collision course. To be sure, there are some workarounds. For coastal regions, there is large-scale desaliniza-tion. But that solution is, once again, costly. Yet letting Los Angeles and San Francisco wither away would be an even more expensive prospect. For inland cities such as Las Vegas and Phoenix, the prospects are more daunting, especially as we factor in the steadily decreasing snowpack-fed river and stream flows, a key source of freshwater for human consumption and irrigation throughout the American West. Even the less-arid Pacific Northwest, with its populous cities Portland, Seattle, and Vancouver, isn't immune—it, too, is starting to face significant water-resource pressures.

Meltwater from mountain glaciers and snowpack is a major source of freshwater in arid and semiarid regions around the world. As we lose Ernest Hemingway's "snows of Kilimanjaro" (the *ice fields* of Kiliman-jaro, to be more scientifically precise), so too do we lose a key source of year-round freshwater for the people of eastern equatorial Africa. Glacial meltwater from the Himalayas feeds the Ganges, Indus, Mekong, Yangtze, and Yellow Rivers, which provide drinking water and irrigation for large populations throughout China, India, and other Asian nations. As we lose the Himalayan glaciers, we lose a primary source of freshwater for more than 1 billion people. A similar threat holds for many other regions around the world.

Then there is groundwater and the aquifers that feed it, accounting for as much as 40 percent of all drinking water and widely used for agriculture in semiarid regions. The Ogallala Aquifer, which lies below the Great Plains of the United States, is among the world's largest aquifers, occupying nearly 200,000 square miles (322,000 square kilometers). It provides nearly 33 percent of the groundwater used for irrigation in the United States and more than 80 percent of the drinking water for the approximately 2 million inhabitants of the region. Unfortunately, the Ogallala Aquifer—like many aquifers—is living on borrowed time. It is being depleted by several percent per decade, with the total depletion to date exceeding the annual flow of eighteen Colorado Rivers. Some estimates have it being completely depleted by 2028.[8] The aquifer consists largely of relict meltwater from the end of the most recent ice age. Once it is depleted, it will take more than 6,000 years to replenish naturally through rainfall (which amounts to about 0.1 inch [0.254 centimeter] per year).

The Food–Water–Energy Nexus

Food, water, and energy are now inextricably linked. The choices we make about one have profound consequences for the others.

Consider again the Ogallala Aquifer, which provides drinking water in the semiarid High Plains of Texas. Water from the aquifer is being tapped not only for irrigation of crops but also for hydraulic fracturing ("fracking")—a process used to free natural gas trapped in bedrock. This is potentially problematic on several levels. First, it means that in some communities, people are forced to take the costly measure of trucking in drinking water because agricultural and fossil fuel interests are competing for the diminishing available groundwater. Adding insult to injury, the mix of toxic chemicals used in fracking can potentially contaminate the groundwater that remains.

Consider also the controversial Keystone XL pipeline, a project to deliver low-quality, heavy oil (bitumen) extracted from the Athabasca oil sands in Alberta to the global market by way of the Great Plains. The pipeline would travel through a large region underlain by the Ogallala Aquifer. If the pipeline were to erupt, its contents could potentially contaminate that critical supply of freshwater. Although the project is currently on hold after being rejected by the Obama administration, there is no guarantee against this scheme or similar efforts moving forward in the future.

Water is also needed for other sources of energy. Obviously, it is necessary for hydropower and hydrothermal energy generation. But a steady source of running cool water is also needed as a coolant for the production of nuclear energy, hence the location of nuclear power plants near rivers (for example, Three Mile Island is located on the Susquehanna River). Climate change is a major source of vulnerability here because it could lead to reduced or more intermittent stream flows.[9]

The same stream flows are needed to cool coal-fired power plants. The continued burning of coal—the most carbon-intensive form of fossil fuel burning—is, of course, a substantial contributor to climate change. And climate change is, as we now know, reducing river flows (or making them more intermittent) in many regions, which, ironically, is adverse to the continued operation of coal-fired power plants. Perhaps we have identified here one key *negative* (that is, stabilizing) climate change feedback loop?

Now, back to the food part of the nexus. Obviously, we have a finite amount of land available for agriculture and livestock. Food production currently utilizes roughly half of Earth's total land surface area and is rapidly consuming what fertile land remains. If we allow an increasing share of the arable land to be used for growing biofuels such as switchgrass, then we are trading off agriculture for energy. Indeed, the trade-off is even more obvious when we consider growing crops such as corn (which could potentially be feeding the world's hungry and malnourished) for the fuel (ethanol) they can provide. The "food for energy" road is an ethically perilous one, and we appear to be proceeding down it headlong. We are reminded that there is no free lunch, figuratively or literally, when it comes to the trade-offs in the food–water–energy nexus.

Land

We have already seen the trade-offs in how we use the land available to us to meet our diverse needs for water, food, and energy. With a growing global population to feed, there is more pressure to find new land to cultivate for agriculture and livestock while providing enough space for the same growing population. These pressures lead to greater amounts of deforestation, which itself contributes nontrivially to our net carbon emissions, further amplifying climate change. Another feedback loop!

Climate change presents a further challenge in this case because it is reducing the available living space, especially in the most populous regions. More than 33 percent of all people live within 60 miles (100 kilometers) of the coastline. Roughly 10 percent live in the zone that lies within 30 feet (9 meters) of elevation above sea level. Herein lies the rub: that zone is highly vulnerable to the combined effects of sea-level rise and more intense hurricanes.

The highest documented storm surge in the United States happened in August 2005 with strong category 5 Hurricane Katrina, one of the most intense hurricanes on record during the most devastating Atlantic hurricane season on record. The town of Pass Christian, Mississippi, recorded a storm surge just less than 28 feet (8.5 meters). Add a couple of feet of sea-level rise (expected within a few decades), and you've got your 30-foot (9-meter) storm surge. Calculations show that many locations are *already* subject to storm surges well higher than 30 feet. A theoretical peak storm surge of 34 feet (10 meters) near John F. Kennedy International Airport in New York City is predicted for just a category 4 hurricane landfall.[10]

Combine projected sea-level rise (as much as 6 feet [2 meters] by the year 2100 and potentially far more a century later) with the stronger storm surges produced by intensified hurricanes, and you get—forgive the pun—*a perfect storm*. You get Superstorm Sandy. But not just once in a century. Fueled by warming oceans and boosted by sea-level rise, Sandy-like strikes on New York City will occur every few years.[11] No wonder the mayor of New York City has created a commission to help find a way to mitigate the coming storm—of climate change.

The United States is among the ten countries in the world with the most people living in a vulnerable coastal zone. The others are China, India, Bangladesh, Vietnam, Indonesia, Japan, Egypt, Thailand, and the Philippines, many of which have very large, impoverished populations that will be forced to flee their increasingly uninhabitable low-elevation coastal environment.

Demographic trends aren't helping matters. Two-thirds of our mega-cities (for our purposes, cities with more than 5 million people) are at least partially in these lowland areas. And more people are moving to these cities.

Other regions, such as the Sahel of Africa, are becoming uninhabitable for climate-related reasons. It is estimated that 80 million of the 100 million people living in the Sahel are dependent on rainfall for their survival. Sustained drought and the growing intermittency and unpredictability of rainfall, however, have rendered the region unfavorable for farming and herding. Villagers are left with no choice but to abandon their land in search of more favorable lands.[12]

There is a name for this phenomenon: *environmental refugeeism*. The potential conflict between the environmental refugees and the native inhabitants of the lands to which they are fleeing is a very potent ingredient in the mix of factors leading national security experts to fear a future global conflict crisis.[13]

Health

By 2030, it is estimated, climate change will cause as many as 700,000 additional deaths a year worldwide.[14] By comparison, each year 443,000 people currently die prematurely from smoking or exposure to secondhand smoke. One could well argue, from this standpoint, that the industry-funded campaign to deny the effects of human-caused climate change has and will cost even more lives and constitutes an even greater crime against humanity than the tobacco industry's campaign to deny the health effects of tobacco.

Malnutrition kills more than 7 million people a year, many of them children. More than 2 million people a year die from complications arising from lack of access to clean drinking water, such as diarrhea and waterborne diseases. The vast majority, again, are children. The adverse impacts of climate change on food and water will magnify the fatalities. The developing world, with its weak health-care infrastructure, will be least able to cope.

A warmer Earth means more extreme, dangerous heat and more fatalities from heat stroke and heat exhaustion. In the United States, there has been a doubling of record daily high temperatures in the past half-century, and this increase has taken a toll. As many as 10,000 people perished in the Chicago heat wave of 1988, which affected a very large portion of those most vulnerable: the elderly and infants. Fortunately, with increasingly widespread air-conditioning in homes, buildings, and vehicles, we in the United States are insulated from the full impacts of heat extremes, and fatalities from complications related to intense heat are limited to fewer than 1,000 a year. (But, of course, the down side of air-conditioning is that it requires large amounts of electricity, which in turn require the burning of more fossil fuels.)

Other countries that have less of this infrastructure aren't as fortunate. The record heat wave in Europe in 2003 took a toll of 70,000 human lives, and the record heat in Russia in 2010 took another 56,000. The heat wave in India and Pakistan in 2015 claimed several thousand more. The elderly, the very young, outdoor workers, and those who lack access to shelter are most vulnerable.

Then there are the impacts of stronger, more devastating tropical storms. Spiked by record ocean temperatures that can be tied at least in part to climate change, Typhoon Haiyan became in November 2013 the strongest tropical cyclone ever to make landfall.[15] All told, 10,000 perished in its wake. The record Atlantic hurricane season in 2005, fueled by unusually warm Atlantic Ocean temperatures, saw an unprecedented twenty-eight named storms and a record four category 5 monster storms. Among them was Katrina, which destroyed much of New Orleans when it made landfall. Nearly 4,000 lives had been taken by the time that season ended.

It is possible to look at catastrophic weather events and think that they are just random, unconnected episodes. But they are not. The heat wave, drought, and massive fire outbreaks in Russia in the summer of 2010 were a manifestation of the same unusual large-scale atmospheric pattern that led to exceptionally strong monsoons and severe flooding in South Asia and India. Pakistan was especially hard hit, with 1,600 people dead and 2 million homeless. That pattern was likely favored by warming oceans.[16]

In fact, more devastating floods, more frequent and expansive wild-fires, and more persistent heat waves and droughts are potentially favored by the impact that climate change is having on the jet stream. There is evidence that the warming of the Arctic may be causing the Northern Hemisphere jet stream to be less active and more likely to get locked into place for weeks on end. When this happens, you get more persistent, unusual weather patterns, like those we have seen in recent years.[17]

Climate change brings not only death but pestilence, too. We can expect to see infectious diseases such as Dengue fever and malaria spread into the extratropics as the globe continues to warm. West Nile virus was first detected in New York City following the record warm year of 1998, and dangerous Hantavirus appears to be spreading north in the western United States.

Then there's the issue of air quality, allergies, and asthma. Higher atmospheric CO_2 favors weeds such as ragweed, whose pollen triggers allergies and worsens asthma. Rising temperatures increase ground-level ozone smog, which also worsens asthma. The number of pollen allergy and asthma sufferers appears to be increasing in recent decades as the globe continues to warm.

Ecosystems

Polar bears and penguins *do* have a role in this discussion, for they symbolize the threat. A world without the polar bear or emperor penguin, a world without an ice-capped Mount Kilimanjaro or a Great Barrier Reef, is a world that has lost some its majesty and wonder. Is that the sort of world we want to leave behind for our children, grandchildren, and great-grandchildren?

But the polar bears and penguins will hardly be alone as they exit the planet. Climate change (and ocean acidification from increasing CO_2 levels) could herald the sixth major extinction event in geological history.

The critics will point to the fact that CO_2 levels were even higher when the dinosaurs roamed the planet, and life clearly thrived then. One encounters that argument in various incarnations. Former president George W. Bush's NASA administrator, Michael Griffin, for example, once asserted that it was "arrogant" for anyone to believe that the "particular climate that we have right here today . . . is the best climate for all other human beings."[18]

Such arguments are, of course, obtuse, for it is the *rate* of warming, not the warmth itself, that poses a threat. Over millions of years, animals and plants can migrate. Coral reefs can drift poleward or equatorward as seas warm or cool. Corals, oysters, and other shell-forming critters aren't endangered by slow changes in atmospheric CO_2, which don't cause the spikes in ocean acidity created by rapid CO_2 increases.

Animals can even evolve—like the polar bear did when it split off from the brown bear as the Arctic ice cap expanded hundreds of thousands of years ago. Although new species may arise in the future, that won't happen anywhere near fast enough to replace the extinctions. We must face the fact that a sixth mass extinction would impoverish the world *now*. An era of new species, if such an era can ever come to pass, might be a million years away.

We are now asking plants and animals to migrate at unprecedented rates and with brand-new obstacles such as cities and highways standing in their path. Polar bears have nowhere farther north to go as the Arctic sea ice disappears. The adorable pikas that live at the tree line of the Rockies have nowhere to go as the treeline gives way to alpine forest.

And frogs that inhabit tropical cloud forests have nowhere to go as global warming literally lifts the cloud base off the mountaintop.

As the seasonal cycles in temperature and rainfall shift, altering by different amounts the timing of the hatching of insects and the arrival of birds, entire food webs are in danger of disruption. Plants and animals possess a certain amount of behavioral elasticity, but the more rapid the changes, the more likely this intrinsic adaptive capacity will be exceeded, and the more likely that we humans will be responsible for one of the most devastating extinction events in Earth's history.

Economy

Critics sometimes claim that it will "cost too much" to do something about climate change. The truth is that it will cost too much *not* to do something about it. We have already seen the damaging impacts that climate change is having on our lives—on food, water, land, energy, our health, and the health of our ecosystems.

But how do we estimate the cost of losing unique ecosystems? How do we put a price tag on the polar bear? How do we calculate the loss of the Great Barrier Reef?

The value of a livable planet is its replacement cost. And since we currently know of no other planet that can support life, the cost of replacing Earth is infinite. So just about any monetized estimate of the cost of damages from climate change is going to undervalue the true cost. There are many "hidden costs" of our current reliance on fossil fuels, so-called externalities—damage done to us and the environment on which we depend that is not accounted for in the current market economy (certainly not without a price on carbon).

There is the damage to the Gulf of Mexico done by the Deepwater Horizon oil-rig disaster in April 2010, the true extent of the ecological destruction only beginning to be understood. There is the potential danger posed to our supplies of drinking water and to our health by the leakage

of chemicals into groundwater from fracking. There is the destruction of mountaintops (and the entire surrounding environment) to get at the coal contained below.

And let's not forget the foreign wars we are fighting to keep oil flowing from dangerous regions of the world such as the Middle East. Think of these wars as a $100 billion subsidy to the fossil fuel industry, courtesy of you and us, the American taxpayers.

Recognizing that *any* estimate is therefore going to be too low, let's nonetheless look at some of the numbers.

There is the destruction of our coastlines, towns, homes, and personal property. Damage by weather events made more extreme by climate change amounts to billions of dollars a year in the United States alone. The United States has sustained more than 170 weather and climate disasters (droughts, floods, tornados, hurricanes, and wildfires) since 1980, the cost of each of which exceeded $1 billion. The total cost exceeded $1 trillion.[19] Globally, extreme weather events in 2013 led to a record forty-one such billion-dollar weather catastrophes. Superstorm Sandy alone did $65 billion of damage. Even ignoring the other impacts that climate change may have had on this event, the indisputable aggravating factor of global sea-level rise led to $2 billion of additional damage in that storm alone.[20]

The direct health-care costs alone are estimated to rise to as much as $4 billion a year within the next decade and a half due to more widespread infectious disease, heat stress and exhaustion, and other health ailments made worse by climate change. This figure doesn't even factor in the losses to our economy from a diminished, less-healthy workforce.

As temperatures warm beyond the comfort level for California wine grapes, a $50 billion wine industry could be decimated, shrinking by as much as 70 percent by midcentury. As winter snowpack continues to diminish in the Sierras, Rockies, and other popular winter-sports regions, the $12 billion ski industry in the United States has already seen losses exceeding $1 billion.[21]

Then there is the onslaught of extreme weather tied in one way or another to climate change. The North American drought and heat wave in the summer of 2012 made a direct strike on America's breadbasket, causing widespread failure of corn, sorghum, and soybean crops. All told, $31 billion of damage was done. The ongoing California drought has cost billions more.

But we live in a global marketplace, where damaged crops in one location and disruption of agricultural supply chains lead to spikes in food prices far afield. The failed harvests in Queensland during the extreme floods of 2010 and 2011 led to a 30 percent increase in food prices across Australia. The severe ongoing drought in Africa has caused a 393 percent rise in sorghum prices in Somalia and a 191 percent rise in corn prices in Ethiopia. The ongoing drought in California has spiked the price of almonds by 50 percent and the price of romaine lettuce by 37 percent in the United States.

The same abnormal jet-stream pattern in 2010 that produced extreme heat and persistent wildfires in Russia and heavy flooding in South Asia and Pakistan wreaked havoc on international food prices. The drought and fires in Russia and Ukraine destroyed much of the wheat harvest that year, leading to as much as an 80 percent increase in global wheat prices. The extreme monsoon rainfalls in Thailand and Vietnam raised rice prices by 25 to 30 percent throughout Southeast Asia.[22]

When all these economic costs are tallied (and, remember, much is neglected in such valuations), it becomes clear that climate change is currently costing the world economy about $1.2 trillion a year, the equivalent of 1.6 percent of total global economic output. That figure could rise to 3.2 percent by 2030.[23]

When economists consider this cost of *inaction* and weigh it against the cost of *taking action*—that is, combating climate change by reducing global carbon emissions—they conclude that *taking action* is a no-brainer. The cost of climate change damages are already greater than the cost of reducing emissions (for example, through a carbon tax or emission-permit system) and will only become more so over time.[24]

And for the critics who say that we shouldn't act as long as there is any uncertainty at all about the impacts of climate change, the economists who have studied this problem say just the opposite. Uncertainty is a reason to act more definitively and more quickly because of the "long tails" of the probability distribution of the impacts of climate change—that is, because of the small but nonzero possibility that the impacts will in fact be far *worse* than we currently predict.[25]

We purchase fire insurance for our homes even though the probability that we will ever experience a house fire is relatively low (less than one in four). By contrast, it is near certain that we will see dangerous and irreversible changes in our climate if we continue with the business-as-usual burning of fossil fuels. Think of climate action as a very wise planetary insurance policy.

It is common to see the figure 450 ppm cited as the maximum "safe" level of CO_2 in the atmosphere because that is the level that will likely prevent warming beyond the 3.6°F (2°C) level often considered to constitute "dangerous" climate change. Neglecting for the time being that it could reasonably be argued that dangerous climate change has *already* arrived, let's scratch a bit beneath the surface of the calculations that are involved. It turns out that the best estimates indicate that limiting CO_2 to 450 ppm yields roughly a 67 percent likelihood of keeping warming below the 3.6°F "dangerous" level.[26]

"What?!" you exclaim. "You're telling us that there's still a 33 percent chance of dangerous warming even if we hold it below the level the scientists are telling us to?"

Yes, that's *exactly* what we're telling you. Dangerous climate change is even more likely than your house catching on fire even *if* we hold CO_2 below 450 ppm—a task that will take a very concerted effort at this point.

In what other circumstances would we accept a 33 percent chance of a catastrophic outcome as acceptable? Yet that's exactly what we're doing when we buy into the conventional framing of what is required to avoid "dangerous" climate change.

Our discussions about climate change too often become hijacked by the physical scientists and economists. But what we're talking about—the health of the planet and the welfare of its current and future inhabitants—is much, much more than a problem of physical science or economics. It is fundamentally an issue of *ethics*.

If we are *stewards* of Earth—as many believe to be our mandate—then it is our responsibility to preserve our planet, not mortgage it for the short-term gain of narrow vested interests. Pope Francis stated in his encyclical on climate change in 2015:

> A threat to peace arises from the greedy exploitation of environmental resources. Monopolizing of lands, deforestation, the appropriation of water, inadequate agro-toxics are some of the evils that tear man from the land of his birth. Climate change, the loss of biodiversity and deforestation are already showing their devastating effects in the great cataclysms we witness. . . . The establishment of an international climate change treaty is a grave ethical and moral responsibility.

The pope admonished those who would ignore his warning: "God will judge you on whether you cared for Earth."[27]

Although technically religious, this credo is arguably in fact independent of one's faith (or lack thereof) and is instead steeped in basic ethical principles. "When the spiritual leader of a billion people says climate change is a moral question . . . then all of us who have been saying that for years are strengthened," commented one observer.[28]

Consider, too, the comments by geophysicist Marcia McNutt, editor in chief of the leading scientific journal *Science*, in an editorial published just weeks after the pope gave this encyclical: "I wonder where in the nine circles Dante would place all of us who are borrowing against this Earth in the name of economic growth, accumulating an environmental debt by burning fossil fuels, the consequences of which will be left for our children and grandchildren to bear?"[29]

Scientists typically do their best to avoid weighing in on sociopolitical debates. When they do, it is generally because they feel compelled to do so by the gravity of the circumstances. In the case of climate change, the situation couldn't be more dire.

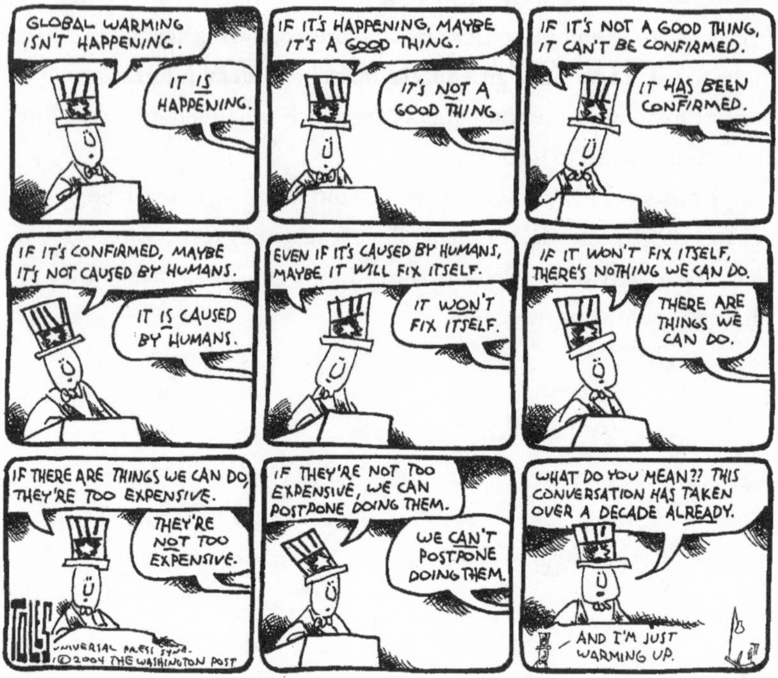
GLOBAL WARMING ISN'T HAPPENING.
IT IS HAPPENING.
IF IT'S HAPPENING, MAYBE IT'S A GOOD THING.
IT'S NOT A GOOD THING.
IF IT'S NOT A GOOD THING, IT CAN'T BE CONFIRMED.
IT HAS BEEN CONFIRMED.
IF IT'S CONFIRMED, MAYBE IT'S NOT CAUSED BY HUMANS.
IT IS CAUSED BY HUMANS.
EVEN IF IT'S CAUSED BY HUMANS, MAYBE IT WILL FIX ITSELF.
IT WON'T FIX ITSELF.
IF IT WON'T FIX ITSELF, THERE'S NOTHING WE CAN DO.
THERE ARE THINGS WE CAN DO.
IF THERE ARE THINGS WE CAN DO, THEY'RE TOO EXPENSIVE.
THEY'RE NOT TOO EXPENSIVE.
TOLES
UNIVERSAL PRESS SYND.
©2004 THE WASHINGTON POST
IF THEY'RE NOT TOO EXPENSIVE, WE CAN POSTPONE DOING THEM.
WE CAN'T POSTPONE DOING THEM.
WHAT DO YOU MEAN?? THIS CONVERSATION HAS TAKEN OVER A DECADE ALREADY.
AND I'M JUST WARMING UP.

1·2·04

[4]

THE STAGES OF DENIAL

Given the overwhelming strength of the evidence, as we have seen, that climate change is (1) real, (2) caused by humans, and (3) a grave threat, one might rightfully ask how it is that some of our most prominent elected officials can still deny that climate change is even happening.

The answer, of course, is that climate change denial isn't really about the science; it is instead about the politics. It is about powerful vested interests that find the implications of the science (that there is a need to stop burning fossil fuels) inconvenient. It is about a massive disinformation campaign to justify an agenda of inaction. Denial that a problem exists in the face of overwhelming scientific consensus that it does leads to often amusing verbal contortionism as climate-denying politicians and talking heads navigate their way through the "six stages of denial."

"It's Not Happening!"

The first stage of denial is rejection of the evidence that climate change is happening. This most evergreen of the denialist canon takes various forms. The most severe is the outright denial that atmospheric CO_2 is even increasing. Although one might think that the overwhelming evidence from direct measurements spanning more than half a century painstakingly documenting the buildup in CO_2 in the atmosphere would appease the doubters, self-styled "skeptics" can still find denier-friendly venues willing to publish denunciations of even this most fundamental of observations.[1]

More often, however, one encounters instead the claim that the globe isn't warming. In the 1990s, climate change contrarians pointed to one apparent discrepancy in the observations: satellite-based estimates of atmospheric temperatures produced by a pair of researchers from the University of Alabama at Birmingham. The data produced by a Microwave Sounding Unit (MSU) appeared to show no warming of the lower atmosphere, apparently contradicting the warming recorded in surface observations. For more than a decade, critics pointed to the MSU data set as evidence that the globe is not in fact warming and celebrated the two scientists, John Christy and Roy Spencer, who themselves were climate change critics.

Although problems with Christy and Spencer's procedures had been suspected for some time, it wasn't until 2005 that other scientists were able to independently analyze the MSU data and expose the glaring errors. Among other problems, there was a sign error (that is, a minus sign where there is supposed to be a plus sign) in their algorithm that turned warming into cooling. The mystery was finally solved. There never *was* any cooling—that conclusion was an artifact of a botched analysis. In the meantime, however, fossil fuel–industry groups and politicians and front groups had used this "evidence" in their bid to justify an agenda of climate inaction for more than a decade.[2]

But never underestimate the persistence of climate change deniers. Even though the faults in Christy and Spencer's work had been exposed, the "globe is not warming" zombie was resurrected in a different guise shortly thereafter. By 2008, it became apparent to climate contrarians that if they cherry-picked their starting and ending dates very carefully and used very short segments of time, they could make it look as if there was no significant rise in temperature.[3] The sleight of hand involves, for example, starting the trend line with the year 1998—an unusually warm year,

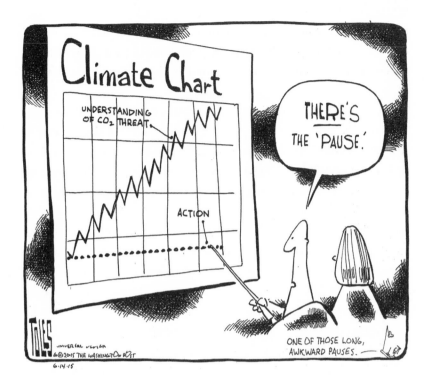

boosted by a major El Niño event—and calculating the trend line over, say, a very short ten-year or so interval (which minimizes the statistical significance). It's like arguing that spring isn't arriving this year because March 27 was warmer than April 9.

Through a psychological phenomenon known as *seepage*, this dubious framing ultimately was adopted by mainstream climate researchers themselves, who would refer to a supposed "pause" or "hiatus" in global warming.[4] Dozens of scientific articles bought into the idea that global warming had, at least temporarily, stopped. The vaunted Intergovernmental Panel on Climate Change (IPCC) even dropped, at least slightly, its lower-bound estimate on Earth's climate sensitivity.[5]

In reality, there was at most only a fleeting slowdown due to natural fluctuations in climate,[6] and the supposed pause or hiatus in global warming collapsed like a house of cards when a new all-time global temperature record was set in 2014 and then again in 2015. In the meantime, however, the "pause" or "hiatus" meme provided ample fodder for climate change deniers looking to forestall policy action.

"OK, It's Happening . . . but It's *Natural*!"

As unprecedented extreme weather and climate events continue to accumulate, climate change deniers have tended to move on to another stage of denial: "Yes, it's happening . . . but it's natural." There are numerous versions of this talking point.

A favorite variant involves pointing to past warm climates that predate any human influence. For example, there is the early Cretaceous period, when dinosaurs roamed the planet and temperatures were higher than they are today. Unless you subscribe to the Creation Museum's account of natural history, human beings weren't around then to emit carbon into the atmosphere.

Or how about the so-called Medieval Warm Period, when temperatures were warmer than they are today? Clearly, there weren't any SUVs around then! Well, first of all, temperatures *weren't* warmer. The best available evidence indicates that although some regions were warm, others were rather cold, and if you average the temperatures over the globe at the time, temperatures did not approach what they are today. But don't let those facts get in the way of this denialist talking point! (Indeed, the reason climate change deniers so fiercely attacked the hockey stick curve is that it refuted this precious, if rather disingenuous, argument.)[7]

There is an even more fundamental fallacy at work here, however . . . so fundamental, in fact, that it even has a name: *post hoc, ergo propter hoc* (after this, therefore because of this). The fallacy is that because some factor may have been responsible for changes in the past, any changes today must be caused by that very same factor. It is akin to arguing that because wildfires happen naturally, the suspected arsonist found with matches and lighter fluid isn't guilty of starting the raging wildfire and should walk free.

In the case of the so-called Medieval Warm Period, natural factors, including a slight uptick in solar heating and a dearth of volcanic eruptions, led to the moderation of temperatures from the eleventh to fifteenth centuries in comparison with the cooler Little Ice Age conditions of the seventeenth to nineteenth centuries. The very same natural factors instead favored *cooling* over the past century. The fact that the globe has warmed *in spite of* those factors can be explained only by human-caused increases in greenhouse gases. As for the warmth of the early Cretaceous period, it is hardly cause for comfort. That warmth was caused by the very same mechanism causing warming today: high CO_2 levels.

Here's the difference, though: those high ancient CO_2 levels resulted from very slow geological processes. These processes subsequently buried much of the carbon beneath Earth's surface over the course of the

past 100 million years. Today, through the extraction and burning of fossil fuels, we are releasing that carbon back into the atmosphere over a time frame of 100 years, a million times faster than nature was able to bury it. As Sting, former lead singer and bass player of the Police, once said, the mighty *Brontosaurus* may have a message for us. Will we heed that message?

"The Problem Is Self-Correcting Anyway"

Faced with overwhelming evidence that we cannot explain the warming of the past century or the array of associated changes in our climate from natural factors, climate change contrarians will often move on to the next stage of denial: "OK, the globe is warming, and, no, it cannot be explained by natural factors, but the changes will be modest. There are *self-correcting* mechanisms, after all, that will kick in and limit the warming and its impacts."

Consider atmospheric scientist Richard Lindzen of MIT. Lindzen has some impressive credentials, having been elected to the National Academy of Sciences for some of his early research. But he is also an equal-opportunity science denier who disputes the health threat from tobacco products[8] and has long argued that stabilizing mechanisms (called "negative feedbacks") will keep warming to a minimum.

Recall our earlier discussion of feedbacks. There are both *positive* feedbacks, which amplify any initial warming (think *vicious cycle*), and *negative* feedbacks, which diminish any initial warming (think *self-correcting mechanisms*). The overwhelming consensus is that the positive feedbacks, on balance, win out.

If you double CO_2 concentrations, the direct increase in temperature from the greenhouse effect of the CO_2 alone is about 1.8°F (1°C)—no problem, right? But the warming results in more water vapor in the atmosphere (the water-vapor feedback). Water vapor is an even more potent greenhouse gas than CO_2 and adds 2.7°F (1.5°C) of warming. The melting of ice leads to increased absorption of sunlight by Earth's surface (the ice-albedo feedback), adding another 0.9°F (0.5°C). It's these positive feedbacks that turn what would be a modest warming (1.8°F) warming into a potentially catastrophic one (5.5°F [3°C]).

Now, some additional potentially important feedback mechanisms are less certain. Arguably chief among them are so-called *carbon-cycle feedbacks*. The oceans and plants have so far been able to absorb about half of the CO_2 we have emitted into the atmosphere. They are a so-called *sink* of carbon. But that *sink* may soon saturate and could even turn into a *source*. If that were to happen, then CO_2 would start to accumulate more rapidly in the atmosphere, even without any increase in emissions.

Consider also methane. Although methane exists in the atmosphere in far lower concentrations than CO_2 (it is measured in parts per *billion* rather than parts per *million*), it is an even more potent greenhouse gas. It is possible that substantial amounts of methane currently trapped in the slowly thawing Arctic permafrost or along coastal shelves in the form of what is known as clathrate could be released into the atmosphere, leading to significant additional warming. Indeed, there is evidence in the geological record that such mechanisms may have been behind past catastrophic warming events.

The scientific community's attention has increasingly turned to the potential threat of these less-certain but potentially dangerous positive feedbacks. Yet climate change contrarians remain utterly preoccupied with the possibility of neglected *negative* feedbacks.

Lindzen, it seems, has never met a negative feedback he didn't like. In 1990, he argued that global warming would lead to a drying of the upper atmosphere. Since water vapor is a greenhouse gas, that drying would lead to less warming—that is, a negative feedback. Lindzen ultimately gave up arguing for that mechanism after it was discredited by the work of other scientists.[9] He subsequently turned his attention to *clouds*. Clouds, after all, are complicated. Different cloud properties can act as both a positive feedback and a negative feedback.

On the one hand, high wispy cirrus clouds warm the surface by allowing solar heat to pass through but blocking the escape of heat from the

surface back out to space (similar, in some respects, to the atmospheric greenhouse effect). On the other hand, other types of clouds, especially thick, low-level stratus clouds, actually cool the surface by reflecting sunlight back out to space (much in the same way that snow and ice do). So knowing whether cloud properties act as a positive or a negative feedback depends not only on knowing whether there are fewer or more clouds in a warmer world but also, just as important, on knowing what types of clouds they will be.

The prevailing view is that the warming and cooling effects of clouds nearly balance and that the net cloud feedback is somewhere between a weak negative and a weak positive. As we learn more, however, the consensus seems to be moving increasingly toward the view that clouds constitute a positive feedback; that is, they exacerbate the warming.[10]

Lindzen has argued otherwise, however. In 2000, he argued for an overlooked negative cloud feedback. In what he referred to as the "Iris Hypothesis," he claimed that global warming would lead to fewer high clouds, thus helping cool the surface. That claim has not held up: other scientists have independently investigated and rejected this mechanism based on both observational and conceptual grounds.

Undaunted, Lindzen argued in 2009 for yet a different negative cloud feedback, claiming that global warming would lead to more of the reflective low clouds. That claim, too, withered under other scientists' scrutiny.[11] Lindzen's quest presumably continues today.

The mechanisms that Lindzen proposed were at least interesting and arguably led to some worthwhile science as other scientists delved into the underlying atmospheric physics to either confirm or, as it turned out, reject his claims. It is difficult to be as charitable, however, when it comes to other proposed "self-correcting" climate mechanisms.

Consider, for example, a study touted a few years ago by James Taylor—no, not "Sweet Baby James" but a public-relations professional of the same name who works for the Heartland Institute, which denies both climate change and tobacco's ill effects on health. The study in question was coauthored by Roy Spencer—one of the two climate change contrarians behind the satellite-based temperature record that supposedly contradicted evidence of surface warming—and his colleague William Braswell.

Taylor's press release about the study proclaimed that the blockbuster new study "indicates far less future global warming will occur than . . . models have predicted . . . and [that] increases in carbon dioxide trap far less heat than alarmists have claimed."[12] So was this study the death knell of human-caused global warming?

Alas, no. Upon publication, the article was thoroughly picked apart by experts in the field for employing an unacceptably simplistic model of the climate system (one is reminded of Einstein's famous quote that things "should be made as simple as possible, but not simpler") and for giving completely different results if any data set other than the one the authors had curiously selected was fed into their model. The editor in chief of the journal that published the study was so troubled that the peer-review process had failed to identify the "fundamentally flawed" nature of the study that he resigned in protest.[13] Although perhaps not going down in flames in such a spectacular manner, most other theories of magical climate "self-correction" have fared little better.

"And It Will Be *Good* for Us!"

After conceding that climate change is happening, is caused by us, and is not completely minimal, deniers often retreat to the position that the effects won't be bad. In fact, they may even be *good*.

"Polar bears are thriving!" they say. But the facts say otherwise: despite the increase in the numbers in some populations of bears due to tighter hunting restrictions, of the twelve populations that have been tracked, eight are decreasing in number, three are stable, and only one is increasing. Polar bears, as a consequence, are classed as vulnerable by the International Union for Conservation of Nature and Natural Resources and are listed as threatened under the Endangered Species Act.[14]

"Agriculture will flourish from higher CO_2 levels (plants just *love* CO_2) and longer growing seasons!" say the contrarians. But the actual science shows that productivity of crops will plummet in the tropics with even modest warming. The devastating impact that extreme weather fueled by climate change has already had on crops across the globe in recent years, as we have seen, paints an even bleaker picture.

Sea-level rise? No problem! "Over the past two years, sea levels haven't increased at all—actually, they show a slight drop. Should we not be told that this is much better than expected?" asks the Polyannish, self-styled "skeptical environmentalist" Bjorn Lomborg (who is neither a skeptic nor an environmentalist in the true sense of those words).[15] In this case, Lomborg was pointing to a short blip in the record related to the El Niño phenomenon.

In reality, not only has sea level continued its inexorable rise, but that rise is accelerating upward over time.[16] This particular episode of denial

fits a larger pattern wherein the so-called skeptical environmentalist has systematically misrepresented the scientific evidence so as to downplay the threat of climate change and to justify an agenda of inaction.[17]

But so what if the Greenland Ice Sheet does melt, anyway? After all, that would expose a lush new continent ripe for human colonization. Why do you think the Norse settlers named it Greenland in the first place? It's because the warm conditions that prevailed during the "Medieval Warm Period," 1,000 years ago when the Norse first settled the continent, made it a prosperous, *green land*, beckoning settlers from far and wide, until the demise of human settlements centuries later with the chill of the Little Ice Age.

Except that's entirely wrong. Accounts like this, common in climate denialist lore,[18] ignore the facts that (1) Greenland was not warmer at the time of Norse settlement than it is today; (2) it has been almost entirely covered in ice for more than 100,000 years—the Viking settlements were restricted to a relatively ice-free fringe (the warmer southern fjords) that is inhabitable today; (3) naming the continent Greenland was, in essence, a marketing strategy to lure settlers from mainland Europe, not an accurate description of this ice-covered continent; and (4) if the Greenland Ice Sheet were to melt in its entirety, far more land would be lost to global sea-level rise than gained from newly uncovered ground in Greenland.[19]

One prominent contrarian, Roger Pielke Jr.—a political scientist at the University of Colorado who publishes on climate policy—continues to insist that climate change has had no detectable impact on damages related to extreme weather.[20] He makes this claim despite assertions to the contrary by such reputable institutions as the Red Cross,[21] Lloyd's of London, and researchers working for leading reinsurance companies Munich Re and Swiss Re.[22] Pielke, who has been invited to testify before Congress by Republicans dismissive of climate change,[23] employs an unusual methodology that involves dividing damages by global economic growth. Leading experts have pointed out that this inappropriate procedure likely removes the very signal of damage by climate change that is being sought.[24]

If you see a pattern here, it might be because there is one. Many of the protagonists advocating against urgent and concerted climate action employ a straw-man framing of climate change impacts as either minimal or beneficial. That may not be an outright rejection of the fundamental physical mechanisms of climate change, but it is still a form of denial, and it provides a convenient fallback position for polluting interests as they find it increasingly difficult to dismiss climate change outright.

"It's Too Late or Too Expensive to Act . . . and We'll Find Some Simple Technofix Anyway"

And finally, we have reached the final stage of denial: "It's too expensive to do anything about it anyway," or, alternatively, "We'll find some cheap technofix at some point."

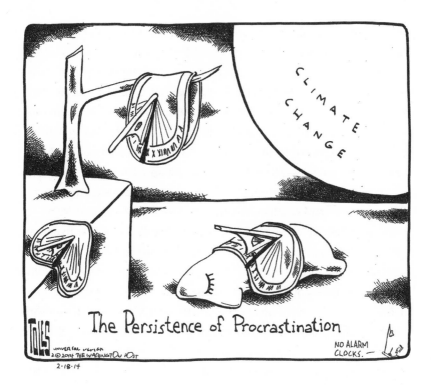

The Persistence of Procrastination

NO ALARM CLOCKS. —

This position, too, is propped up by myths and fallacies. Chief among them is that inaction is the least-expensive path forward. As we have seen, that view is simply wrong. Economists looking at the problem have determined that the cost of doing nothing is far greater than the cost of taking action. That point can't be emphasized enough because the critics love to focus on the costs on the *action* side while completely ignoring the greater costs on the *impact* side.

Another prevalent fallacy used to bolster this position is that "you can't chew gum and walk at the same time" or, to be more specific, that acting on climate change will somehow divert resources from our efforts to address other pressing problems. One particularly egregious form of this argument involves the invented concept of "energy poverty" pushed by folks such as Rex Tillerson, CEO of ExxonMobile, who once posed, without any apparent sense of irony, the question "What good is it to save the planet if humanity suffers?"[25]

The "energy poverty" conceit is also promoted by "skeptical environmentalist" Bjorn Lomborg, former Microsoft CEO Bill Gates,[26] and the Breakthrough Institute. The last declares its mission to be looking for a "breakthrough" to solve the climate change problem, but at bottom appears to be opposed to anything—be it a price on carbon or incentives for renewable energy—that would have a meaningful impact. Meanwhile, it remains curiously preoccupied with opposing advocates for meaningful climate action[27] and is coincidentally linked to natural gas interests.[28]

The idea behind "energy poverty" is, first, that lack of access to energy (rather than to, say, food, water, health, and so on) is the primary threat to people in the developing world. If you're willing to accept that dubious premise, then consider another. The argument then goes that fossil fuels are the only viable way to provide that energy. Got it? If you care about the disadvantaged of the world, then you should be bullish on fossil fuels.

Someone ought to tell that to Pope Francis, a well-known voice for the poor. He has specifically rejected the "energy poverty" myth, explaining how distributed, renewable energy in the form of solar power and hydropower is far more practical in most of the developing world.[29] And he's not alone in making this assertion. Even the fossil fuel–friendly *Wall Street Journal* recently noted that "renewable energy could offer a . . . solution for remote areas, because it is created and consumed in the same region and doesn't require massive power plants and hundreds of kilometers of power lines."[30] If you've lost the *Wall Street Journal* . . .

There is an even deeper problem with the premise that climate action detracts from other concerns. Climate change, as we have seen, actually *exacerbates* other societal challenges—food, water, health, land. Pope Francis emphasized this simple fact in his recent encyclical on the environment, and the Department of Defense takes this position as well.[31] The nail in the coffin of the "energy poverty" myth is that the impacts of climate change will actually place far more people *in* poverty. Don't take our word for it, though. According to a World Bank study reported by generally denialist Fox News, climate change could "thrust 100 million into deep poverty by 2030."[32]

Finally, what about those cheap technofixes that have been proposed? They include, among other fanciful schemes, placing giant mirrors in space to reflect back sunlight, shooting huge amounts of particulate matter into the upper atmosphere to mimic volcanic eruptions, and pouring large quantities of iron into the oceans to "fertilize" them. We cannot do justice to the topic of "geoengineering" in a paragraph or two, so we have devoted a whole chapter to it (chapter 7).

We have seen all sorts of excuses for inaction, from (1) the globe isn't warming to (2) it is, but it's natural; (3) it's caused by humans, but it's minimal; (4) it'll be good for us anyway; (5) it's too expensive to do anything about it; and, finally, (6) we'll find a simple or inexpensive technological fix. Each excuse is a form of denial and should be recognized as such. There is no simple way out. Ultimately, we're left with one real solution: reducing our collective carbon footprint.

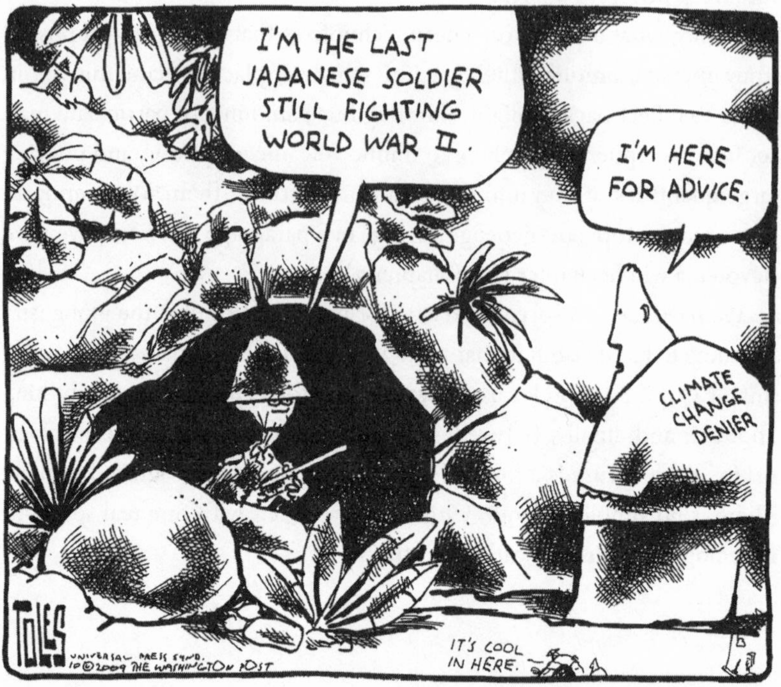
I'M THE LAST JAPANESE SOLDIER STILL FIGHTING WORLD WAR II.
I'M HERE FOR ADVICE.
CLIMATE CHANGE DENIER
IT'S COOL IN HERE.
TOLES
UNIVERSAL PRESS SYND.
10 © 2009 THE WASHINGTON POST

10.29.09

[5]

THE WAR ON CLIMATE SCIENCE

Like the sole Japanese soldier found still fighting World War II in the 1970s,[1] the war on climate science may well continue as long as there are fossil fuels to be mined and mercenaries to be hired. But it is important to understand how the current assault fits into a longer, historical pattern.

For decades, industry special interests have financed expensive publicity campaigns, provided extensive funding to think tanks and front groups, and hired scientists with impressive credentials to attack scientific findings that threaten their product—be it tobacco, industrial chemicals, or fossil fuels.[2]

The war on science can be traced back more than half a century, beginning with the activities of the tobacco industry in the 1950s. The industry knew by then that it had a problem on its hands. Its own internal research revealed that its product was harming and killing people. In his book *Doubt Is Their Product*, the title of which is taken from a tobacco industry internal memo, Assistant Secretary of Labor David Michaels explains how the tobacco industry failed to take ownership of the problem. It instead "manufactured uncertainty by questioning every study, dissecting every method, and disputing every conclusion" and in the end "successfully delayed regulation and victim compensation for decades."[3]

Books such as Chris Mooney's *Republican War on Science* and Naomi Oreskes and Erik Conway's *Merchants of Doubt* describe how the methods and modus operandi developed by the tobacco industry in the 1950s and 1960s were finely honed in subsequent decades as other industry special interests adopted and adapted them in an effort to confuse the public and policy makers about the threats posed by their products.[4]

Cutting Their Teeth

The "tobacco strategy" has repeatedly been deployed in efforts to discredit scientific evidence of harm to the environment.[5] Among its earliest adopters was the chemical industry in its campaign to vilify Rachel Carson in the early 1960s. Carson's book *Silent Spring*, often credited with ushering in the modern environmental movement, brought widespread awareness of the harmful environmental impacts of the pesticide DDT (dichloro-diphenyltrichloroethane).[6] The president of Monsanto Corporation, the largest producer of DDT, denounced her at the time as "a fanatic defender of the cult of the balance of nature."[7]

In the critics' revisionist history, this environmental hero was instead an evil mass murderer.[8] "Millions of people around the world suffer the painful and often deadly effects of malaria because one person sounded a false alarm ... that person is Rachel Carson," states the website rachelwaswrong.org, maintained by the industry front group known as the Competitive Enterprise Institute (CEI).[9] Ironically, DDT was phased out not because of the environmental damages exposed by Carson but because it had steadily lost its effectiveness as mosquitoes grew resistant to it.

Next up was acid rain. Evidence mounted by the late 1960s that emissions from midwestern coal smokestacks (sulfur-containing particulates known as "sulfate aerosols," to be specific) were causing problems of their own. The particulates were getting carried aloft by the westerly winds and winding up in rainfall in downwind regions in the northeastern United States. The acid-spiked rainfall was killing off forests and rendering lake and streams lifeless throughout New York and New England.

By the 1970s, the scientific community well understood the mechanisms behind acid rain.[10] And in 1970, President Richard Nixon both established the Environmental Protection Agency (EPA) and signed the Clean Air Act, which granted the EPA the mandate to regulate industrial emissions. Yet as late as the mid-1980s, S. Fred Singer, a physicist whom we discuss more fully later, teamed up with conservative media outlets such as the *National Review* to dispute any human role in the acid rain problem, blaming it instead on natural factors such as volcanoes.[11] This industry opposition and manufactured controversy were adequate for the time being to prevent further policy action.

In 1990, however, George H. W. Bush—Ronald Reagan's vice president and now president—led by a proactive, pro-environment EPA administrator named William K. Reilly, went on to sign into law an innovative, market-based "cap-and-trade" system, thus putting in place financial incentives for polluters to cap their emissions. The effort, which was opposed by industry groups with noble-sounding names such as Citizens for a Sound Economy (Who isn't for a *sound economy*, after all? But in actuality it is a front group, now known as FreedomWorks, founded by Charles and David Koch to block governmental regulations),[12] was a dramatic success. Forests, streams, and lakes have largely recovered over the past several decades.

Then there was ozone depletion. By the 1980s, solid scientific evidence had amassed that chlorofluorocarbons (CFCs), widely used at the time as a propellant in spray cans and as a refrigerant, were destroying the stratospheric ozone layer. The chemical industry, however, predictably attacked the science. The chairman of DuPont dismissed the scientific evidence as "a science fiction tale . . . a load of rubbish . . . utter nonsense."[13] Perhaps his dismissiveness had something to do with the fact that CFCs were an $8 billion market. Singer again played a key role in aiding and abetting industry special interests by contesting the role of CFCs in ozone depletion.[14]

The discovery of the hole in the ozone layer over Antarctica in 1985 and the recognition that the problem was even worse than scientists had predicted nonetheless led to passage of the Montreal Protocol in 1987, which banned industrial use of ozone-depleting CFCs. Despite the continued industry opposition, the United States readily became a signatory under President Ronald Reagan.

You might be forgiven for being just a little confused at this point. Yes: *Republican* presidents Richard Nixon, Ronald Reagan, and George H. W. Bush all supported *regulatory solutions* to emerging global environmental problems despite considerable pushback from industrial special interests. Bush introduced our country to *cap and trade,* a market-driven approach to reducing pollution that is roundly derided by present-day Republicans. The simple fact is that environmental protection wasn't always the partisan political issue that it has become.

Climate Change: You've Got Next!

Well, you can pretty much guess what happened next. As climate change rose to prominence in the early 1990s, climate science would increasingly find itself in the crosshairs. Well oiled now for having fought multiple battles in the preceding decades savaging the science of DDT, acid rain, and ozone depletion, the attack machine was now humming.

Funded by conservative foundations and industry special interests, various think tanks and front groups emerged on the scene, tasked with the role of attacking science that proved inconvenient to their interests.[15] But there was something else in play, something that would have deep implications for the climate change debate.

As Oreskes and Conway note in *Merchants of Doubt,* an entire generation of Cold War–era physicists was primed with an ideologically driven distrust of any limits on freedom. These "free-market fundamentalists" played willfully into the agenda of regulation-averse special interests.[16]

Several of them came together to form the George C. Marshall Institute (GMI)—an industry-funded think tank that would go on to become, as *Newsweek* put it, a "central cog in the denial machine."[17] Formed at the height of the nuclear arms race in 1984, this group focused on defending Reagan's controversial Strategic Defense Initiative, the missile defense system commonly referred to as "Star Wars," against influential critics such as scientist and public figure Carl Sagan.

Sagan and others had argued that the Strategic Defense Initiative would lead to an escalation of the nuclear arms buildup. The massive detonation of a large enough nuclear arsenal in the course of an all-out nuclear war between the United States and the Soviet Union would create enough dust and debris to block out a sufficient amount of sunlight to induce a state of perpetual winter, a "nuclear winter." Such a scenario would be as devastating for humanity as the asteroid-induced global dust storm was for the dinosaurs some 65 million years ago.

Cold War hawks, however, saw nuclear winter merely as a scare tactic employed by peaceniks who had been duped by the Soviets or were sympathetic to their cause. The idea was in itself a threat to our security,

they claimed. GMI therefore sought to discredit the case for concern, going directly after the underlying science. This attack on science involved holding congressional briefings and soliciting popular articles and op-eds to debunk the science. It even included intimidating television networks that considered running a documentary on nuclear winter.[18]

Here's where things get even more interesting and even more germane to our topic. The projections of nuclear winter were based on early-generation global *climate models*. So if you didn't like the science of nuclear winter, you *really* weren't going to like the science of climate change. Needless to say, climate change would soon attract the attention of GMI and the Cold War physicists it had recruited, who would join forces with conservative foundations and fossil fuel–industry groups to form a formidable coalition to attack the science of climate change.

Deniers for Hire

Frederick Seitz was a world-class physicist with impeccable scientific credentials. He had once served as president of the National Academy of Sciences—the preeminent scientific body in the United States—and as the president of Rockefeller University. He was awarded the coveted Presidential Medal of Science in 1973 "for his contributions to the modern quantum theory of the solid state of matter";[19] the Department of Defense Distinguished Service Award; the National Aeronautics and Space Administration's Distinguished Public Service Award; and the Compton Award, the highest honor of the American Institute of Physics.

Seitz was also, it turns out, a founding figure in the art of modern-day science denialism. In the early 1990s, he went on to chair the GMI full time, where, joined by two other like-minded physicists—Robert Jastrow, founder of the NASA Goddard Institute for Space Studies (now, ironically, a world-class center for climate modeling), and William Nierenberg, one-time director of the hallowed Scripps Institution for Oceanography—he campaigned against the science of global warming, acid rain, and ozone depletion.[20]

How is it possible, you ask, that three such distinguished scientists joined together in an attack against science? In a word . . . well, *two* words: *ideology* and *money*.

Seitz and his fellow Cold War hawks were opposed to government regulation as a matter of principle. According to Robert Proctor, professor of the history of science at Stanford University and author of *Golden Holocaust: How Cigarette Makers Engineered a Global Health Catastrophe,*

> These free-market fundamentalists, steeped in Cold War oppositions (market economies versus command economies, the individual versus the state, the free world versus Big Brother), attacked any and all efforts to trace environmental maladies back to corporate chemicals. Chlorinated fluorocarbons were not really eating away at the ozone layer, and the sulfates being belched from coal-fired plants were not causing forest-harming acid rain; even secondhand cigarette smoke was not causing any provable harm.[21]

As for money, the famous Upton Sinclair quip "It is difficult to get a man to understand something, when his salary depends on his not understanding it" is once again relevant. After retiring from academia in the late 1970s, Seitz received more than $500,000 from tobacco giant R. J. Reynolds for his advocacy of downplaying the health threats of tobacco. In his subsequent role as chair of GMI, he would receive funding from other corporate interests, including fossil fuel giant ExxonMobil, to downplay the threat of climate change.[22] Seitz was the first of the all-purpose deniers-for-hire. Scientific authority is a valuable commodity when it comes to industry public relations, and Seitz and others like him trafficked in it readily and effectively.

In 1998, Seitz lent his imprimatur to the Global Warming Petition Project (Oregon Petition), opposing the Kyoto Protocol to limit greenhouse gas emissions. The petition was sent out to a broad mailing list nationwide and was accompanied by a cover letter signed by Seitz and using his past affiliation as president of the National Academy of Sciences to urge recipients to add their names to the petition. Also included in the materials was what Seitz referred to in the letter as a "twelve page review of information" about global warming.[23]

In reality, that review was a fake journal article, "Environmental Effects of Increased Atmospheric Carbon Dioxide," by Arthur B. Robinson, Noah E. Robinson, and Willie Soon, formatted to look as if it had been published in the prestigious journal *Proceedings of the National Academy of Sciences.* Smelling a rat, the National Academy of Sciences itself took the unprecedented action of publicly denouncing Seitz's efforts as a deliberate deception, noting that its official position on the science was very different from that expressed by Seitz in his letter.[24]

And what about the "Oregon Petition" itself? With its impressive-sounding 31,000 nominal "scientist" signatories, it is still touted by climate change deniers as supposed evidence of widespread doubt among scientists about human-caused climate change. However, an analysis by *Scientific American* found that few of the signatories were scientists, and a number of those who were indeed scientists were deceased. Among the putative signatories were Geri Halliwell (yes, she of the Spice Girls) and B. J. Hunnicutt, a fictitious character from the television series *Mash.* You get the picture.[25]

S. FRED SINGER
"Carbon dioxide is not a pollutant. On the contrary, it makes crops and forests grow faster."

If Seitz was the *first* of the all-purpose deniers-for-hire, S. Fred Singer was and is the most *prolific.* The parallels between Singer and Seitz are in fact striking. Singer, too, was a Cold War physicist (a rocket scientist—literally) and an academic (a faculty member in the Department of Environmental Sciences at the

University of Virginia). Like Seitz, Singer left academia (in 1990) to found a think tank, the Science and Environmental Policy Project, with a mission of debunking the science of ozone depletion, climate change, tobacco, and sundry other environmental and health threats. He has received considerable funding from corporate interests, including Philip Morris, Monsanto, and Texaco.[26]

Now consider an infamous incident involving the respected scientist Roger Revelle, who is credited for, among other things, having inspired Al Gore's concern about climate change when Gore was a student at Harvard. Revelle made fundamental contributions to our current understanding of human-caused climate change, providing key evidence in the 1950s that the burning of fossil fuels was elevating greenhouse gas concentrations and making some of the earlier estimates of the expected global warming.

Shortly before Revelle passed away in 1991, Singer added Revelle as a coauthor to a paper he had written for the journal *Cosmos*, published by the Cosmos Club, a Washington, D.C., intellectual society. The paper, which was nearly identical to an earlier denialist article by Singer, disputed the evidence that climate change was being caused by humans. Revelle's secretary and a former graduate student have suggested that Revelle was quite uncomfortable with the manuscript and that the denialist framing was added after Revelle, who was gravely ill and died just months after the paper's publication, had any opportunity to see the revised article.[27]

Singer was also the principal behind the infamous report of the so-called Nongovernmental International Panel on Climate Change (NIPCC) in 2008. Funded by the Heartland Institute, an organization that had fronted for the tobacco industry in decades past and that today fronts for fossil fuel interests, and formatted to mimic the authoritative assessment reports of the Intergovernmental Panel on Climate Change (IPCC), the NIPCC report instead sought to undermine the IPCC's findings. It has been dismissed by ABC News as "fabricated nonsense."[28]

Numerous other scientists, supported by fossil fuel interests directly or indirectly to dispute various aspects of the scientific evidence, have fomented climate change denialism.[29] You'll recall Richard Lindzen of MIT, who has argued that hypothetical previously undiscovered stabilizing feedback mechanisms will minimize the warming. There's also the team of John Christy and Roy Spencer, also mentioned in chapter 4, whose erroneous satellite-based estimates of atmospheric temperatures were propped up for more than a decade as evidence against global warming. There is Patrick Michaels, formerly of the University of Virginia and now of the conservative Cato Institute, who has argued that climate change will be either minimal or good for us.

And then there is Willie Soon of the Harvard-Smithsonian Astrophysical Observatory in Cambridge, Massachusetts, who has impressively gone through all the stages of climate change denial, arguing everything from "global warming is natural" to "it's good for the polar bears." It turns out that the articles he was publishing in the peer-reviewed literature were secretly being described as "deliverables" to the fossil fuel interests funding him.[30]

Climate change deniers and contrarians are affiliated with a plethora of organizations and front groups funded by the fossil fuel industry and receive money from them. Among this "Potemkin village" of entities, as Oreskes and Conway have described them,[31] are the Advancement of Sound Science Center, Alexis de Tocqueville Institution, Americans for Prosperity, Cato Institute, CEI, Fraser Institute, FreedomWorks (formerly Citizens for a Sound Economy), GMI, Heartland Institute, Hudson Institute, Media Research Center, National Center for Policy Analysis, and many more.[32]

Orchestrators of climate change denialism have built their Potemkin village, manned it with troops who are ready and willing to defend it, and established a clear mission: "mislead large sections of the American public into thinking that the evidence for human-caused warming [is] uncertain, unsound, politically tainted and unfit to serve as the basis for any kind of political action."[33]

But He Plays One on TV

Recall that the modern-day attack machine got its start with big tobacco in the 1950s. Perhaps it is therefore unsurprising that there is a direct link between those who once advocated for tobacco interests and now advocate for fossil fuel interests. S. Fred Singer, for example, is today funded by various fossil fuel entities to dispute global warming. But back in 1995, he worked for a Philip Morris front group to help brand the threat of secondhand smoke as one of the "the top five environmental myths."[34] Indeed, many of the scientists-for-hire who advocated for tobacco interests in decades past are today working for fossil fuel interests to discredit the science underlying human-caused climate change.[35] They are nothing if not versatile.

Consider Steven J. Milloy, the self-avowed "junk man" who rails against what he calls the "junk science" on DDT, ozone depletion, and all matters of "environmental extremism" on his website junkscience. com and in interviews and op-eds in conservative-leaning media outlets.[36]

Like many of the deniers-for-hire, Milloy got his start in tobacco— Philip Morris, to be precise.[37] But he's become an all-purpose environmental critic. Consider the work he has done for Syngenta, a European agrichemical company that markets pesticides known as neonicotinoids, which have been implicated in colony collapse disorder,[38] the mass die-off of bee populations, including a shocking 42 percent die-off of honeybees in the United States during 2014.[39]

Syngenta also markets a weed killer, atrazine, that has been linked to the die-off of frogs. Milloy has received undisclosed sums of money from Syngenta to be its mouthpiece on atrazine: an e-mail message from Milloy to Syngenta's corporate head of public relations in 2008 read, "Beth Carroll said that you had some Atrazine talking points for me. Would love to see them."[40]

Milloy has also helped target Tyrone Hayes, a researcher at the University of California, Berkeley, who has spent more than a decade researching atrazine and has published work indicating that it causes gender abnormalities in frogs exposed to doses even lower than the EPA limits. Milloy has called Hayes a "Berserkeley atrazine-hater" and has accused him of lying about his science. In a commentary titled "Freaky-Frog Fraud," he stated that "Hayes seems to be determined to scare the public about atrazine."[41]

Milloy is no scientist,[42] but he's darned good at playing one on television—Fox News, to be specific—where he is presented as an environmental science expert. Milloy, in this capacity, regularly calls out the "junk science" implicating tobacco products in human health ailments, pesticides in environmental ailments, and fossil fuels in our current planetary ailment.

What Milloy has failed to disclose while busy presenting himself as an independent "junk science" expert on Fox News is that, as noted earlier, he has accepted payments from Phillip Morris, ExxonMobil, and Syngenta for his advocacy efforts.[43] When journalist Paul Thacker reported these facts in the *New Republic* in 2006, he also reported the reaction from Fox News, which claimed to be unaware of the financial ties and gave Milloy the lightest of slaps on the wrist, conceding only that "any affiliation he had should have been disclosed."[44]

**"JUNK MAN"
STEVEN MILLOY**
"We don't agree . . . that man-made emissions of carbon dioxide (CO_2) and other greenhouse gases are having either detectable or predictable effects on climate."

Milloy is hardly alone. Numerous lawyers, lobbyists, and political operatives have served as mouthpieces for industry special interests. Marc Morano, the original architect of the swift-boat attacks against John Kerry during his presidential bid in 2004, now launches similar smear campaigns against climate scientists. Along with Milloy and Morano, Christopher Horner and Myron Ebell of the fossil-fuel-industry–funded CEI can regularly be seen on Fox News attacking climate science and climate *scientists*.

The Art of the Ad Hominem

Among the more unseemly tactics utilized by the industry-funded disinformation machine is the politics of personal destruction, the ad hominem attack, the smear. It is the last line of defense in a losing battle. Knowing that they don't have the facts on their side, the forces of denial have increasingly turned to personal attack as their weapon of choice. It's all about distracting the public, vilifying the enemy, and going straight at that enemy's strength—in this case, trust and credibility.

Now, as we have already stressed, there is nothing wrong with criticism in science. Indeed, criticism is an important part of the self-correcting machinery of science. But it has its limits. Criticizing someone's science is quite different from attacking him or her personally, engaging in name-calling and false accusations, and issuing personal threats against him or her.

You will recall that the libertarian think tank CEI has accused Rachel Carson, widely regarded as the founder of the modern environmental movement for her work on the harmful effects of DDT, of being a mass murderer. To ensure that she not lie peacefully in her grave, the attack was resurrected in 2012, the fiftieth anniversary of the publication of *Silent Spring*. Among other hit pieces, a commentary at the conservative *Forbes* website, "Rachel Carson's Deadly Fantasies," accused her of "gross misrepresentations," "atrocious" scholarship, and "egregious academic misconduct."[45] Its author was Henry I. Miller, an adjunct fellow at CEI and a scientific advisory board member at GMI, who, unsurprisingly, is widely known for his advocacy for the tobacco industry.[46]

Why the extreme vilification of Carson? It is rather simple. She's an icon, a symbol of the environmental movement. Discredit her personally, and all justification for concern about the environment melts away—or so the thinking goes, anyway. For the same reason, climate change deniers have targeted Al Gore, who in the world of politics has become very closely associated with the climate change issue. Discredit Gore as a person, and all concerns about climate change presumably come tumbling down like a house of cards. So the critics have sought to take down Gore by any means possible. They have criticized his weight, his energy bills, and incidents in his personal life—indeed, pretty much anything else they can scrape up in their efforts to discredit him personally.

Let us now, in this context, revisit the story of Roger Revelle and S. Fred Singer. Recall that Revelle was instrumental in inspiring the

TOLES UNIVERSAL PRESS SYND. 3© 2007 THE WASHINGTON POST

Congress finally settles on a response to global warming

WRONG PLANET—

3·23·07

young Al Gore to become interested in the issue of global warming and that Singer signed Revelle's name to a denialist paper he published in the journal *Cosmos* shortly before Revelle passed away. These two facts are not unrelated.

Revelle's graduate student at the time of the *Cosmos* incident, Justin Lancaster, has stated that Singer "hoodwinked" Revelle into adding his name to the article and that Revelle was "intensely embarrassed that his name was associated" with it. Lancaster characterized Singer's behavior as "unethical" and, furthermore, had a strong suspicion of Singer's intent: to discredit Al Gore and his campaign to raise public awareness about the threat of climate change. Lancaster stands by these charges despite legal threats against him by Singer.[47]

Lancaster's assertion would seem to be confirmed by the manner in which Singer and his fellow denialists for years used Revelle's "coauthorship" of the *Cosmos* article to their advantage, trumpeting the talking point that even "Gore's global-warming mentor" believed that "drastic, precipitous and, especially, unilateral steps to delay the putative greenhouse impacts can cost jobs and prosperity and increase the human costs of global poverty, without being effective."[48]

Soon after Revelle's death, his daughter became fed up with the distortion of her father's views and legacy and penned an op-ed in the *Washington Post* denouncing how "critics of Sen. Al Gore have seized these words to suggest that Revelle, who was also Gore's professor and mentor, renounced his belief in global warming" and stressing that "nothing could be further from the truth." She reaffirmed that Revelle "remained deeply concerned about global warming until his death in July 1991."[49]

Many others have suffered attacks by the smear machine. Consider the distinguished Stanford ecologist Paul Ehrlich. In 1968, Ehrlich, like Rachel Carson, provided an early warning that human consumption was on a collision course with finite planetary resources in his classic book *The Population Bomb*. And like Carson, he was denounced by the usual suspects, with the Cato Institute's Julian Simon calling him "an alarmist purveyor of doom and gloom" who was leading a "juggernaut of environmentalist hysteria."[50] Yet several decades later, a group of more than 1,500 of the world's leading scientists, including half of the living Nobel Prize laureates, confirmed Ehrlich's findings, concluding that "human beings and the natural world are on a collision course" and that human activity is inflicting "harsh and often irreversible damage on the environment and on critical resources."[51] The major national academies of the world have issued similar statements.[52]

The Smear Comes to Climate Science

Climate change deniers have turned the ad hominem attack into an art form. Consider, for example, the case of Stephen Schneider of Stanford

University, one of the first climate scientists to publicize the threat of human-caused climate change.

Schneider was perhaps the most articulate public voice the climate research community has ever produced. He was a premier climate scientist who made fundamental early contributions to our understanding of greenhouse warming in the 1970s and equally fundamental contributions to the science of climate risk assessment in later decades. But he was also a natural-born communicator with skills that rivaled Carl Sagan's. That made him, like Sagan, a threat.

In response, climate change deniers widely misrepresented Schneider's views. There is, for example, the widespread myth that still circulates in contrarian circles that Schneider was predicting another *ice age* in the 1970s, the implication being that this terrible prediction means that we can dismiss his later warnings about global warming. He wasn't, but the truth—that scientists like Schneider were still wrestling with the competing effects of aerosol cooling and greenhouse warming in the early 1970s—is nuanced and easily misrepresented.[53]

Far more insidious, however, is the way that one particular quote by Schneider has been misrepresented by climate change deniers over the years. Here we give the entire quote from an interview he did with *Discover* in October 1989 because it's important to know exactly what he said:

> On the one hand, as scientists we are ethically bound to the scientific method, in effect promising to tell the truth, the whole truth, and nothing but—which means that we must include all the doubts, the caveats, the ifs, ands, and buts. On the other hand, we are not just scientists but human beings as well. And like most people we'd like to see the world a better place, which in this context translates into our working to reduce the risk of potentially disastrous climatic change. *To do that we need to get some broadbased support, to capture the public's imagination.* That, of course, entails getting loads of media coverage. So we have to offer up scary scenarios, make simplified, dramatic statements, and make little

mention of any doubts we might have. This "double ethical bind" we frequently find ourselves in cannot be solved by any formula. Each of us has to decide what the right balance is between being effective and being honest. *I hope that means being both.*[54]

What Schneider was clearly saying was that he hoped would-be communicators would strive both to be accurate and balanced in how they communicate their findings to the public and to be effective communicators at the same time.

But observe how Julian Simon, the Cato Institute smearer of Paul Ehrlich, altered the quote to grossly misrepresent Schneider in a commentary published in the *American Physical Society Newsletter* in March 1996. Simon fabricated an alternative version of Schneider's quote wherein "To do that we need to get some broadbased support" was replaced with "Scientist [*sic*] *should consider stretching the truth* to get some broad base support." Not content with having leveled the false accusation that Schneider had encouraged "stretching the truth," Simon also eliminated the final line, "I hope that means being both," which explicates Schneider's imperative that scientists be *both* honest and effective.[55]

Then there is Ben Santer, a leading climate scientist and member of the National Academy of Sciences. Because of his important scientific contributions to the key conclusion of the IPCC's second assessment in 1995 that there was now a "discernible human influence on climate,"[56] Santer was viewed as a threat by the forces of denial. But, have no fear! The two Freds soon came to the rescue.

First, S. Fred Singer published a letter in the journal *Science*, falsely asserting that inclusion of Santer's work in the report violated IPCC rules.[57] Shortly thereafter, in a commentary on the right-wing editorial pages of the *Wall Street Journal*, Frederick Seitz accused Santer of "scientific cleansing"—a charge that was especially distasteful given that Santer had lost relatives in Nazi Germany.

As Singer and Seitz slowly disappeared from view, others appeared to replace them. Marc Morano got his start working for conservative "shock jock" Rush Limbaugh before moving on to work for the ExxonMobil-funded Conservative News Service (now known as Cybercast News Service [CNS]). In this capacity, he helped launch one of the most disreputable smear campaigns in modern U.S. political history, the so-called swift-boat campaign, designed to take a great strength of then presidential candidate John Kerry—his distinguished military record, including three Purple Hearts earned in Vietnam—and turn it into a political liability via fabricated accusations and innuendo regarding Kerry's patriotism.[58] This tried-and-true method of attack politics may well have been a critical contributor to Kerry's loss in the election.

MARC MORANO
"[Climate scientists] deserve to be publicly flogged."

Morano then applied the same tactics in the climate change arena. As a staff member for the leading climate change denier in the U.S. Senate, James Inhofe (R-Okla.), he once again aimed to take a great strength—the trust that the public has in scientists when it comes to the issue of climate change—and erode it through character attacks and smear campaigns. In other words, he took "swift boating" to climate science.

James Hansen, former head of the NASA Goddard Institute for Space Studies, was the first climate scientist to publicly proclaim, in the sweltering Senate chamber in June 1988, that climate change was upon us, and he has remained an authoritative voice calling for action in the years since. Hansen was therefore considered a threat to fossil fuel interests. But have no fear! Marc Morano to the rescue . . .

In his campaign against Hansen, Morano labeled him a "wannabe Unabomber" who supports "ridding the world of industrial civilization," "razing cities," and "blowing up dams." He suggested that Hansen is mentally unbalanced (rhetorically asking whether it's "time for meds"), all because of his call to regulate carbon emissions.[59]

One of us (Michael Mann) has also been at the receiving end of Morano's attacks, being called a "charlatan" responsible for "the best science that politics can manufacture."[60] Compared with vilification by the *Wall Street Journal* and Fox News, witch hunts by fossil fuel–funded politicians, accusations of fraud, and comparisons with convicted child molesters in a tag-team attack by the CEI and the *National Review*, Morano's critique frankly feels a bit pedestrian.[61]

Truth Must Prevail

Human activity has always been in some sense an ongoing experiment. Progress is built on these experiments. But our experimentation can occasionally cause large unforeseen problems, and we sometimes find out the hard way. Toxins, radiation, and drug side effects are just a few examples.

Burning fossil fuels turns out to be another example. Subtle, initially invisible, but, as it turns out, extremely consequential. Everybody understands the immediate dangers of a poisonous gas such as carbon monoxide, and we have taken steps to control it. Carbon dioxide is a different type of danger. Carbon monoxide is an immediate and acute threat to human health. Carbon dioxide is a long-term threat to global climate stability.

But the path to understanding and dealing with these threats is fundamentally the same: through science. And so we develop an understanding of the effects of the pollutants, measure them, and determine what is necessary to minimize the dangers.

This is not rocket science, but it is science. What has happened in the case of CO_2, however, is that some very powerful interests don't like the prescription offered by our planetary doctors.

Their response has been to attack not only the message, but also—as we have seen—the messenger. That's unfortunate for two fundamental reasons. First, it has postponed a rational, timely response, which has made the threat all the more dire. Second, by attacking the very process of science itself, they have also confused and hamstrung the mechanisms by which we understand and address *any* problem. It is akin to attacking and undermining the structure of the English language to the point where communication between people can no longer take place.

This ill-advised mode of response is dangerous for the climate, dangerous for the bedrock practice of science on which our whole technological civilization rests, and dangerous for a fact-based political discourse on which our whole system of democratic government rests.

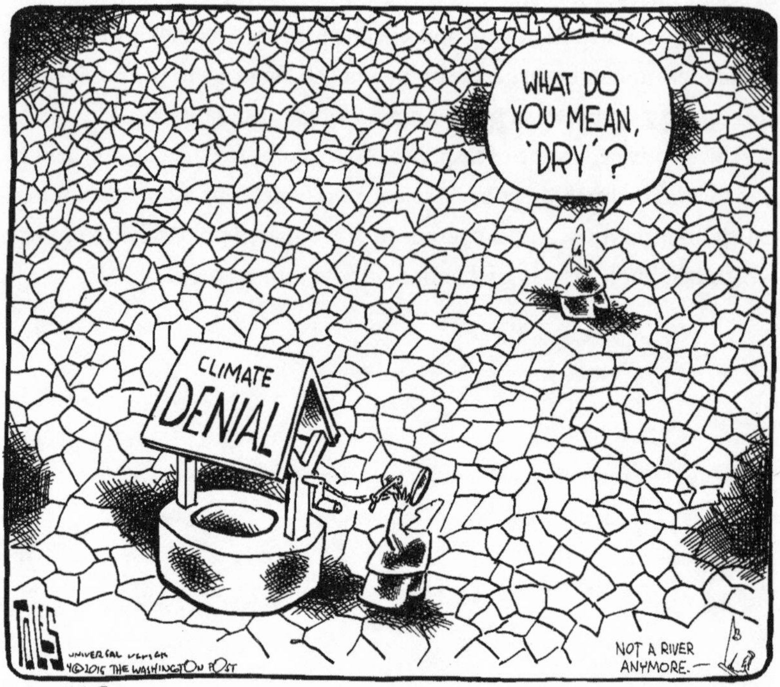

WHAT DO YOU MEAN, 'DRY'?
CLIMATE DENIAL
TOLES
©2015 THE WASHINGTON POST
NOT A RIVER ANYMORE.—

4·13·15

[6]

HYPOCRISY—THY NAME IS CLIMATE CHANGE DENIAL

If you are searching for hypocrisy, look no further than the public discourse over human-caused climate change. Among the corporate special interests that fund organizations and front groups to attack the science and confuse the public; the talking heads, public-relations mercenaries, and politicians who serve as willing accomplices; and the agenda-driven media outlets that serve as mouthpieces for their propaganda, there is fertile ground for ridicule.

Heads Buried in the Sand

The best examples of hypocrisy can, of course, be found in the words and actions of politicians who deny climate change. Many have quite literally buried their heads in the sand when it comes to the threat of climate change.

One is reminded, for example, of how Republican presidential hopeful Mitt Romney mocked President Barack Obama's concern over the "rise of the oceans" in the closing weeks of the presidential campaign in 2012— unusually bad timing, coming just weeks before the calamitous landfall of Superstorm Sandy. President Obama and Governor Chris Christie (R-N.J.) together toured the swath of coastal devastation along the iconic New Jersey shore in a bipartisan display of fortitude and commitment as the election neared. Global sea-level rise added 25 square miles (65 square kilometers) and at least $2 billion to that devastation.

And we can move on to Virginia. Norfolk is home to the world's largest naval base, a major contributor to the state's economy. Experiencing recurring floods when storm surges combine with high tide and sea levels that rise 1 foot (30 centimeters) every five decades, the entire Hampton Roads region is already threatened by climate change.

When legislators from Hampton Roads in the Virginia General Assembly requested state funding in 2012 to study the potential impacts of sea-level rise, however, "Tea Party Republicans" cried foul. The terms *climate change* and *sea-level rise*, they felt, were "liberal code words." Only when these words were replaced with the "politically neutral" phrase *recurrent flooding* did state Republicans, including the climate change–denying governor, Bob McDonnell, approve the study.[1]

Then there is Virginia's former attorney general Ken Cuccinelli. Cuccinelli, also a "Tea Party Republican," attempted to sue the University of Virginia a few years ago because the government funded one of its professors (coauthor of this book, Michael Mann) to study the issue of climate change (of course, nearly all research in the natural sciences is funded by the government).

Issuing a so-called civil investigative demand—a civil subpoena originally designed to ferret out Medicare fraud (reasoning that climate change research itself constitutes a fraud)—Cuccinelli demanded all the professor's personal e-mail exchanges with more than thirty climate scientists around the world. Major scientific and academic groups and organizations throughout the country insisted that the University of Virginia stand up to what they viewed as a transparent effort by Cuccinelli to intimidate a researcher whose findings had proved inconvenient to Cuccinelli's funders.

The university subsequently hired outside counsel and challenged the legality of the subpoena. Prominent media outlets such as the *Washington Post* and its cartoonist, Tom Toles (the other coauthor of this book), covered the affair, criticizing and mocking Cuccinelli's "witch hunt."[2]

Cuccinelli lost in the lower court, which rejected his case: the court found that in his forty-plus-page filing, Cuccinelli had failed to provide *any evidence of wrongdoing.* So Cuccinelli took the matter to the Virginia Supreme Court, which ultimately reaffirmed the lower court's finding and rejected the case with prejudice. In other words, the justices really didn't want to see this sort of attempted abuse of power ever again.[3]

Brushing off the defeat, Cuccinelli made a run for governor in Virginia in 2013. The scientist whom Cuccinelli had attacked played a prominent role in the campaign, writing op-eds, doing interviews, and participating in television campaign advertisements for Cuccinelli's opponent, Terry McAuliffe. He joined McAuliffe on a three-day "science week" campaign tour in early July and appeared onstage with McAuliffe and former president Bill Clinton after introducing Clinton at a campaign rally in early September.

Antagonizing Thomas Jefferson's university, as Cuccinelli had, apparently didn't sit well with Virginia voters. McAuliffe was victorious. Cuccinelli, now without a job, instead opened an Oyster farm on Tangier Island—an island in Chesapeake Bay slowly being inundated by global sea-level rise.[4]

Never let anyone tell you that elections don't matter. While in office in 2010, Governor Bob McDonnell (who was later convicted of corruption charges and sentenced to prison),[5] had attempted to bury climate as an issue by disbanding Virginia's Commission on Climate Change.[6] Among Democrat Terry McAuliffe's first acts as governor in 2014 was to revive the commission. Michael Mann was invited to be a member.

If you thought that Virginia's problem is an isolated one, think again. North Carolina, its immediate neighbor to the south, has suffered a remarkably similar antiscience fervor of late. Republicans in the state recently attempted to outlaw climate model predictions of accelerating sea-level rise.[7] One half-expects them to follow up by denying entry to land-falling Atlantic hurricanes, especially those with Hispanic names.

When it comes to head-in-the-sand burial, though, it is Florida that wins—hands down. With 1,200 miles (1,932 kilometers) of coastline and more than 5 million residents who would be displaced by just 10 feet (3 meters) of sea-level rise—something we are likely now committed to, though how quickly it will happen is still rather uncertain[8]—Florida arguably has more to lose by unmitigated climate change than any other state. What is Republican governor Rick Scott's plan for responding to the threat? How about banning the use of the terms *climate change* and *global warming* in all official state communications and publications?[9]

Florida's junior U.S. senator and former presidential hopeful, Marco Rubio, rates little better. His approach to dealing with climate change is to attack the scientists, deny any human role in global warming, and oppose all viable policy solutions.[10]

Denialism Knows No (State) Boundaries

Climate denialism isn't confined to the coastal states. By some measures, in fact, it runs even stronger as one moves inland. Consider Senator James Inhofe (R-Okla.), a recipient of extensive funding over the years from fossil fuel interests, including ExxonMobil and the Koch brothers, Charles and David.[11] Inhofe is perhaps best known for declaring that climate change is the "greatest hoax ever perpetrated on the American people"[12] and for introducing a snowball on the floor of the Senate as ostensible proof that global warming isn't happening.

While presiding as chair of the Senate Committee on Environment and Public Works, Inhofe has held a number of hearings over the years that attempt to debunk the science of climate change. In 2003, one of us (Michael Mann) had the distinct pleasure of testifying at one of those hearings, alongside two industry-funded climate change deniers (one of whom was Willie Soon, mentioned in chapter 5).[13]

More memorable, however, was a hearing that Inhofe held two years later, in September 2005.[14] The hearing wasn't particularly memorable for the shopworn myths that Inhofe and his witnesses parroted, including the

claims that (1) scientists had been predicting an imminent ice age in the 1970s (they had not), (2) major scientific findings haven't been independently replicated (they have—in spades), and (3) climate can't be predicted because weather is unpredictable beyond a week or so (which is tantamount to asserting that we can't predict the seasons because we don't know whether it will snow on January 15 next year).

No, the boilerplate denialist talking points weren't particularly interesting or memorable. What *was* memorable was Inhofe's choice of witnesses. One was retired hurricane expert Bill Gray. Beloved by his many students over the years and respected for his early contributions to the field of tropical meteorology,[15] Gray was also known for his rather cringeworthy forays into areas well outside his expertise—including climate change.[16] But Gray had far more credibility about climate science than the other witness Inhofe invited: Michael Crichton, the well-known author of science-fiction books such as *Jurassic Park*.

Crichton had recently written the novel *State of Fear*—a thinly veiled denialist polemic masquerading as an action-adventure novel—which clearly endeared Crichton to Inhofe. It seemed like a tacit admission of intellectual bankruptcy for Inhofe to turn to a science-*fiction* writer— someone best known for proposing that dinosaurs could be reanimated using frog DNA—to make his case. That concession was seemingly betrayed when Inhofe stated at the conclusion of the hearing, without any sense of irony, that when it comes to climate change, his preference is to "sit back and look at [it] in a non-scientific way."[17]

In July 2011, Inhofe was selected as keynote speaker at the Heartland Institute's annual conference dedicated to denying global warming. He had to cancel at the last minute, however, because he had grown ill swimming in a

lake in his home state, Oklahoma: the lake was suffering from an algal bloom as a result of the unprecedented heat and drought that Oklahoma was experiencing that summer—an event that scientists have tied to climate change.[18]

Finally, there is Inhofe's ideological ally in the House of Representatives, Joe Barton of neighboring Texas. What Inhofe is to the Senate, Barton is to the House. Also a recipient over the years of considerable largesse from the fossil fuel industry, Barton was former chair of the House Committee on Energy and Commerce. His environmental record was so famously dismal that it earned him the nickname "Smokey Joe."

Barton has proved especially proficient at climate change denial. His greatest hits include remarking that "the science is not settled, and . . . is actually going the other way," "we may in fact be going into a cooling period," and "CO_2 is rising, but it is not necessarily . . . causing temperature to rise."[19] In 2006, he commissioned a now thoroughly discredited report attacking the hockey stick curve.[20]

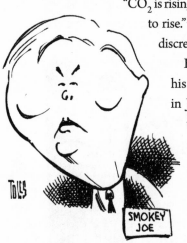

"SMOKEY JOE" BARTON
"The science is not settled, and the science is actually going the other way. . . . We may in fact be going into a cooling period."

But what Joe Barton is most famous for today is his infamous apology to British Petroleum (BP) in July 2010. He felt bad that others had sought to hold BP accountable for the immeasurable environmental damage done by the massive amount of crude oil the company had spilled into the Gulf of Mexico in the Deepwater Horizon oil-rig disaster in April.[21] One of us (Tom Toles) rightly speculated at the time that Barton presumably would have wanted to apologize to King George III for the "lack of gratitude the colonists showed."[22]

Over the course of the following year, Texas lost 25 percent of its cattle and saw its agriculture industry decimated by an unprecedented drought fueled by climate change. One wonders if Smokey Joe has suffered any second thoughts about climate change despite his romantic attachment to the fossil fuel industry.

There is indeed a striking disconnect between the environmental reality that Texas is already dealing with, suffering some of the most devastating impacts of climate change, and the unusually virulent strain of climate change denialism that afflicts so many of its elected officials: Joe Barton, Rick Perry, and Ted Cruz. Take Cruz. Perhaps to curry favor with leading donors such as the Koch brothers during a particularly vital stretch of the Republican primary campaign in 2015, Cruz used his position as chair of the Senate Subcommittee on Space, Science, and Competitiveness to attack the science of climate change as the Climate Change Conference in Paris was drawing to a close.

In December 2015, Cruz held a Senate hearing titled *Data or Dogma? Promoting Open Inquiry in the Debate over the Magnitude of Human Impact on Earth's Climate*,[23] which provided a high-profile forum for advancing climate change deniers' shopworn accusations that (1) the globe isn't actually warming (yes, it is), (2) climate scientists are skewing the data (no, they're not), and (3) the climate research community is censoring contrarian viewpoints (the scientific community abhors bad science but celebrates legitimate novel ideas). He invited three climate change contrarians to attack mainstream climate science but allowed only one witness to testify in its defense, a rather striking imbalance given that the true breakdown is a roughly 97 to 99 percent agreement among scientists that climate change is real and caused by humans.[24] Edward Markey (D-Mass.), the ranking member of the committee, aptly summarized the affair when he noted that "the only thing that requires a serious scientific investigation is why we are holding today's hearing in the first place."[25]

Cruz's efforts, however, pale in comparison with those of Lamar Smith (R-Tex.), chair of the House Committee on Science, Space, and Technology. During his tenure as committee chair, Smith has (1) tried to defund the earth sciences at the National Science Foundation; (2) attempted to redefine the peer-review process at the foundation so that politicians like him, rather than scientific experts, determine what science gets funded; (3) threatened an investigation against a climate scientist who was signatory to an open letter arguing that fossil fuel interests that knowingly hide the dangers of their product from the public should be held accountable in the same way the tobacco industry was; and, most recently, (4) issued a subpoena to leading scientists at the National Oceanographic and Atmospheric Administration demanding their personal e-mails because he disapproved of a study they had published disputing the popular but false contrarian talking point that "global warming has stopped."[26]

It's *Never* the Right Time

When is it the right time to talk about climate change? If you listen to the critics and cynics, apparently *never.*

The Cuyahoga River in Ohio caught fire in 1969. That event was a tipping point in the public consciousness regarding air and water pollution. There was a sudden, widespread awareness of the growing level of pollutants in our air and water, a recognition that it had reached crisis proportions. Within just a few years, we saw passage of both the Clean Air Act and the Clean Water Act. The people spoke, and policy makers listened.

When it comes to climate change, haven't we had our Cuyahoga moment already? Didn't Hurricane Katrina in 2005 adequately awaken us to the growing threat of monster storms? Wasn't the drought in Texas in 2011 a wake-up call for even the reddest parts of red-state America? Weren't the unprecedented heat, drought, and rampant wildfires of the summer of 2012 enough to tell us we have a problem? How about Superstorm Sandy striking New York City and the New Jersey coast in an election year? Still not enough? Then how about an epic drought—the worst in at least 1,200 years—in our most populous state, California? Surely that's enough.

Alas, no single event can be catastrophic enough, it seems, to galvanize public attention and mobilize action. Part of the problem, for certain, is the shrinking American attention span in combination with a modern media environment where few stories, regardless of how important or profound, can persist in the public conversation for more than a few twenty-four-hour news cycles. This atmosphere is prohibitive for raising awareness about slowly growing threats such as climate change.

But that is only part of the problem. A bigger part, arguably, is the concerted effort by special interests and those who do their bidding to ensure that advocates for progress are unable to seize on any potential teaching moments. Whether the topic is Superstorm Sandy or Sandy Hook—that is, the crisis of our climate or the crisis of our schoolchildren's safety—powerful vested interests—fossil fuel interests in one case, the National Rifle Association in the other—are quite happy with the status quo. They simply don't want to see things change. And they readily poison the well whenever there is an opportunity for a teachable moment.

The teaching moments in the case of climate change would seem to be plentiful—droughts, storms, heat waves, and deluges that have been made devastating, unprecedented, extreme, and historic *because of* climate change and provide a glimpse of even worse things to come if we continue our current way of doing things.

These events provide powerful images of what climate change really means and a narrative that communicators can use to explain how climate change is already having an impact on ordinary people, where they work and live, today. Just as with iconic climate change symbols such as the polar bear and the hockey stick, that power makes such events dangerous—dangerous to those banking on public apathy to ensure a policy of continued inaction.

That is why climate change deniers wage their fiercest attacks when it comes to the linkages between climate change and extreme weather. Climate scientists are vilified and mocked when they even suggest the possibility, for example, that Superstorm Sandy's impacts were made worse by climate change (despite the growing body of solid scientific evidence that they were).[27] To use Sandy to talk about climate change, they said, is "to exploit a tragedy," just as they said two months later that talking about gun control in the wake of the Sandy Hook elementary-school tragedy, where more than twenty schoolchildren lost their lives at the hands of a single unhinged gunman with access to semiautomatic weapons, was "exploiting a tragedy." Let us give a name to this cynical gambit—"Sandy silencing."

Naturally, the phenomenon of *seepage* rears its head here. Every time one of these singular events takes place—an extreme event where climate change likely worsened the impact—there is a solid chorus of doubters, credentialed scientists who offer themselves up as "experts" to deny, before the television cameras, that climate change played any role at all. These self-styled "honest brokers" earn themselves a much-desired role in the public conversation while endearing themselves to the critics. A twofer. But they ill serve society at large, which depends on frank and honest assessments of risk from scientists for enlightened policy making.

A sleight of hand is often involved. Take, for example, the record ongoing drought in California. Some scientists have dismissed the possibility of any possible connection with climate change by arguing that the atmospheric circulation patterns leading to low amounts of precipitation are not entirely without precedent. But that argument ignores the fact that it is the simultaneous occurrence of record low precipitation *and* record heat that has generated this unprecedented drought—and the simultaneity of these conditions can, indeed, be linked to climate change.[28]

Now, there apparently *is* an exception to the rule that there is no good time to talk about climate change. The one good time to discuss it, as far as the critics are concerned, is when we seem poised for potential progress, when the issue seems to be catching fire—both figuratively and literally. When a glimmer of hope appears in the horizon, that's when the critics want to talk about climate change—but not in the way you might imagine.

Climategate: The Real Story

There was much hope among advocates for climate action heading into the international Climate Change Conference in Copenhagen in December 2009. In fact, some groups even branded the event "Hopenhagen." The science was ever more clear that climate change was not only real but a serious problem. In the wake of events such as Hurricane Katrina, and with the success of Al Gore's book *An Inconvenient Truth* in 2006, there was growing public recognition that it was time to act on the problem. Copenhagen was where we would finally achieve a breakthrough, where the nations of the world would finally join together and confront the existential threat of human-caused climate change. But it was not to be.

Opponents of climate action dug in their heals and orchestrated the most well-organized, cynical, and effective disinformation campaign yet. In an effort to undermine the public's confidence in the scientific basis for concern about climate change, they manufactured a fake "scandal" in the weeks leading up to the summit, seeking to distract the public and the politicians at this critical moment. Even the name attached to the affair— "Climategate"—was the product of a carefully crafted narrative invented and then foisted on an unknowing public through a collaboration among industry front groups, paid attack dogs, and conservative media outlets.[29]

Thousands of e-mails between climate scientists around the world were stolen from a university computer server in Great Britain late that summer. The e-mails were carefully combed through over several months and organized into an easily accessible archive. Individual words and phrases, harmless in context (for example, *trick* as in "trick of the trade"), were plucked from among the many thousands of e-mails and removed from their original context. Having been cherry-picked and taken out of context, they were then used to misrepresent what the scientists were actually communicating to one another. Finally, climate change deniers had the smoking gun—the scientists themselves were admitting that climate change was a massive hoax! The data had been cooked! The climate scientists were conspiring to pull a fast one on the public!

Front groups connected with the Koch brothers and industry-funded critics such as Steve Milloy helped promote these outrageous untruths, while right-wing media outlets—especially those owned by Rupert Murdoch, such as Fox News and the *Wall Street Journal*, as well as somewhat less-sanitary sources such as the *Drudge Report* and Rush Limbaugh—readily served as a megaphone, filling the airwaves and television screens with false allegations, smears, and untruths about climate change.

At least nine different investigations in the United States and the United Kingdom ultimately determined, however, that there had been no impropriety on the part of the scientists, no fudging of data, no attempt to fool the public about the scientific evidence for climate change. In the end, the only wrongdoing that could be found was the criminal theft of the e-mails in the first place—bitter irony, given that the Watergate scandal, the origin of the moniker "Climategate," was about the theft of documents, not their content.

The criminals were never caught, and the trail went cold. In the meantime, climate change deniers milked the fake scandal for all it was worth, attempting to sabotage the already-delicate negotiations at the Copenhagen summit.

The lead Saudi climate change negotiator, Mohammad Al-Sabban, asserted that the stolen e-mails would have a "huge impact" on the negotiations, issuing the stunning statement that "it appears from the details of the scandal that there is no relationship whatsoever between human activities and climate change."[30]

Right-wing politicians in the United States cheered and pounced. Indeed, it was deliciously ironic that James Inhofe, who had infamously dismissed the overwhelming scientific evidence of climate change as "a hoax," wasted no time whatsoever exploiting the *true* hoax that was Climategate. He used the false allegations against climate scientists to call for the criminal investigation of seventeen climate scientists (including Presidential Medal of Science recipient Susan Solomon of MIT and, yes, coauthor of this book Michael Mann).

And, of course, Sarah Palin threw herself into the fray. Two days into the nine-day Copenhagen Climate Change Conference, Palin published an op-ed in the *Washington Post* promoting a laundry list of false allegations against climate scientists based on out-of-context snippets from the stolen e-mails. She asserted, for example, that the stolen e-mails "reveal that leading climate 'experts' deliberately destroyed records."[31] The claim is a

SARAH PALIN
"Climate change is to this century what eugenics was to the last century."

blatant falsehood, and the scare quotes around "experts" add something special to her remark.

Palin did later concede that "a lot of those emails obviously weren't meant for public consumption" and acknowledged that they could be misinterpreted if "taken out of context."[32] Only, she was talking about her *own* e-mails, which had been released in response to a Freedom of Information Act submitted during her time as governor of Alaska.

Unfair and Unbalanced

As we have seen, the media has played a role in facilitating climate change denialism. Part of the problem is media "false balance," a bad habit acquired in Journalism 101 wherein journalists give fringe viewpoints equal footing with mainstream thinking when it comes to politically contentious issues such as climate change. "Just present both sides!" the instruction goes. But this is the lazy way out of arbitrating a dispute between science and antiscience. Not all viewpoints are equal when it comes to matters of science. There are objective truths—Earth isn't flat, evolution is an observable fact, and climate change is real and caused by humans. These things are true, whether you like it or not—and that, to paraphrase science communicator extraordinaire Neil deGrasse Tyson, is what is so great about science.

This endemic flaw of journalistic false balance, however, has been greatly exacerbated by the increasing polarization of our public discourse and the decentralization of our sources of information. Nowhere is this more true than in the impenetrable right-wing echo chamber.

Most political conservatives get their information from news outlets such as Fox News, which brands itself "fair and balanced." A good rule of thumb is that if a network needs to preemptively announce that it is "fair and balanced," it is probably neither. When it comes to climate change, Fox News has constructed an alternative universe where the laws of physics no longer apply, where the greenhouse effect is a myth, and where climate change is a hoax, the product of a massive conspiracy among scientists, who somehow have gotten the polar bears, glaciers, sea levels, superstorms, and megadroughts to play along.

RUPERT MURDOCH
"Climate change has been going on as long as the planet is here, and there will always be a little bit of it. We can't stop it."

In fact, the problem is hardly confined to Fox News. The entire Rupert Murdoch media empire, News Corp., which includes the *Wall Street Journal* and dozens of newspapers around the world, from the *New York Post* in the United States to the *Sun* and the *Times* in the United Kingdom and the *Australian* and *Herald Sun* in Australia, has been spreading climate change denialist misinformation for years.

Murdoch can deny climate change with the best of them, happily sharing nuggets of wisdom such as "Just flying over N Atlantic 300 miles of ice.

Global warming!";[33] "Climate change has been going on as long as the planet is here, and there will always be a little bit of it"; and "If the sea level rises 6 inches . . . we can't mitigate that, we can't stop it. We've just got to stop building vast houses on seashores."[34] But following a close second to Murdoch in News Corp. stock ownership is Kingdom Holding—otherwise known as the Saudi royal family, the world's leading oil barons and the same folks who used the manufactured Climategate scandal to sabotage the Copenhagen Climate Change Conference in 2009.

Maybe it's just a coincidence that Fox News engaged in nonstop coverage of the fake Climategate scandal and that the *Wall Street Journal* ran a half-dozen Climategate-themed op-eds and editorials in the weeks leading up to Copenhagen.

Fox News and the Murdoch media empire, which—to the consternation of science advocates everywhere—has recently been expanded

to include *National Geographic,*[35] are joined by other right-wing news and pseudonews outlets in the United States, such as the *Drudge Report,* the *Washington Times, Breitbart News,* and the *National Review,* as well as by right-wing radio personalities such as Rush Limbaugh and Glenn Beck to spread climate change disinformation and denialism. And they have gotten help—a great deal of help—from a pair of brothers from Kansas.

A Koch and a Smile

Fred C. Koch was an American entrepreneur who made his fortune in oil refining in the early twentieth century. After being sued for patent infringement in the United States, he built oil-distillation plants in the Soviet Union, which had no intellectual-property rights.[36] As Jane Mayer of the *New Yorker* puts it, "Unable to succeed at home, Koch found work in the Soviet Union." Koch proceeded to help the Stalin regime construct fifteen oil refineries. Subsequently, however, as Mayer explains, "Stalin brutally purged several of Koch's Soviet colleagues. Koch was deeply affected by the experience, and regretted his collaboration."[37]

Koch took the fortune he had earned in the Soviet Union back to the United States, where he subsequently founded the Wood River Oil and Refining Company in 1940. It would go on to become Koch Industries, a major American oil-refining interest. Embittered by his experiences in the Soviet Union, Koch grew into a staunch anti-Communist and Cold War hawk. (Where have we heard this story before?) He went on to found the ultraconservative John Birch Society, dedicated to *limited government.*

Fred Koch's sons, Charles and David, inherited both their father's company and his ideology. With antigovernment leanings and a personal stake in the continued exploitation of fossil fuels, the Koch brothers were perfectly positioned to become champions of climate change denialism. And they have lived up to that potential.

Having diversified into a major global conglomerate, Koch Industries is now the second-largest privately held corporation in the United States and, equally important, the largest privately held fossil fuel interest. The Koch brothers are the greatest potential beneficiaries from construction of the Keystone XL pipeline through Canada and the United States, profiting by as much as $100 billion from delivering the dirtiest, most carbon-intensive petroleum to the open market,[38] a scenario that climate scientist James Hansen has declared "game over for the climate,"[39] though the pipeline is now looking somewhat less likely with the rejection of the project (at least for now) by the Obama administration.

The two oil moguls have used their considerable wealth (currently estimated at more than $100 billion)[40] to fund conservative political activities, contributing tens of millions of dollars to front groups, organizations, and political candidates who support their antiregulatory agenda. That agenda, of course, includes attacks on climate scientists, the propagation of climate change disinformation, and opposition to clean-energy policies.

With the advent of "dark money"—thanks to the Supreme Court decision in *Citizens United v. Federal Election Commission*, allowing unlimited campaign spending by corporate entities (a decision that hinged on the votes of two justices, Antonin Scalia and Clarence Thomas, who regularly attended the Koch brothers' annual retreats)[41]—it is impossible to determine just how much the Koch brothers have spent. But the spending that *can* be traced amounts to hundreds of millions of dollars—enough over the past several election cycles to purchase the U.S. Congress for all intents and purposes. The Kochs are planning to spend the better part of $1 billion in the 2016 election. Perhaps that amount can even buy them a president.[42]

The Koch brothers have used their immense financial resources to support groups that advocate delay and inaction on climate change, giving, for example, more than $100 million since 1997 to groups denying climate change.[43] Indeed, most of the residents in the "Potemkin village" of climate change denial have received or currently receive Koch funding. Prominent among them is Americans for Prosperity, which, in addition to spending hundreds of millions of dollars in campaign advertising and get-out-the-vote efforts for climate change–denying politicians over the past several election cycles,[44] sponsored the Hot Air Tour during the 2008 election year with the slogan "Global Warming Alarmism: Lost Jobs, Higher Taxes, Less Freedom." Its website explains its mission: "Americans for Prosperity is working hard to bring you the missing half of the global warming debate. What will the impacts of reactionary legislation be for you, your family and our economy?"[45]

The Kochs fund the Competitive Enterprise Institute (CEI), which, readers may recall, was involved in early attacks on scientists such as Rachel Carson, fronted for big tobacco in the early 1990s, played a critical role in blocking regulation of greenhouse gas emissions under the Bill Clinton and George W. Bush administrations, and was instrumental in helping manufacture the Climategate scandal in 2009.[46] In 2006, CEI ran the widely ridiculed national campaign singing the virtues of fossil fuel carbon emissions: "They Call It Pollution. We Call It Life."[47] And no, you can't make this stuff up.

The list goes on. Readers may recall the Heartland Institute, whose climate change–denying "report" written by the Nongovernmental International Panel on Climate Change (NIPCC) in 2008, was dismissed by ABC News as "fabricated nonsense."[48] In case you may have thought that the Heartland Institute learned its lesson, it didn't: in 2012, it ran a billboard campaign comparing climate scientists to Unabomber Ted Kaczynski and was caught that same year in a secret effort to infiltrate schools with industry-funded denialist propaganda.[49]

Finally, we would be remiss not to mention arguably the most important member of the Koch family, ALEC. The American Legislative Exchange Council is a corporate mill funded by the Koch brothers and other conservative foundations, fossil fuel interests, and assorted corporate entities and groups. Working through ALEC, special interests get sympathetic politicians to draft and help pass state legislature bills favoring their agendas. ALEC bills often seek to undermine environmental regulations, to deny that climate change exists or poses a threat, and to quash efforts to incentivize renewable energy.

More than 100 companies, organizations, and groups have pulled out of ALEC in recent years, citing disapproval of its antienvironmental agenda. In August 2014, Microsoft withdrew support, citing ALEC's lobbying efforts to block the development of renewable energy.[50] Google dropped out of ALEC in September 2014 because it was "just literally lying" about climate change, said Google's CEO Eric Schmidt.[51] Even oil giant BP left in March 2015, citing similar concerns.[52] But the real bombshell was Shell Oil's withdrawal from ALEC in August 2015, explaining that "its stance on climate change is clearly inconsistent with our own."[53]

Nevertheless, the Koch brothers have many weapons at their disposal. There are the two most prominent congressional climate change deniers—Senator James Inhofe and Representative Joe Barton—both leading recipients of the brothers' largesse over the years. The list of Republicans vying for their party's nomination in the 2016 presidential election, moreover, reads like a who's who of climate change–denying Koch favorites:[54] Jeb Bush (R.-Fla.), Ted Cruz (R-Tex.), Rand Paul (R-Ky.), Rick Perry (R-Tex.), and Marco Rubio (R-Fla.) received substantial amounts of Koch money. One Republican candidate who has conspicuously received *no* money from the Kochs is iconoclast Donald Trump. Although himself a climate change denier, Trump has not endeared himself to the Kochs and has dismissed the other candidates as "puppets" of the two brothers.[55]

Governor Chris Christie (R-N.J.), a former candidate for the nomination, *has* expressed concern about climate change in the past. Despite an apparent effort to make amends for this admission by pulling out of the Regional Greenhouse Gas Initiative, an organization of northeastern states committed to reducing greenhouse gas emissions, he remains out of favor with the Koch brothers. Senator Lindsey Graham (R-S.C.), also a former candidate, has taken a strong stance on climate change in recent years and has spoken out against the *Citizens United* decision, allowing unlimited corporate spending on elections. He hasn't received any money from the Koch brothers in eight years.[56]

The Kochs made no secret of their early preference for Governor Scott Walker (R-Wis.).[57] And why not? As observers have noted, "His track record of actively undermining pro-environment programs and policies while supporting the fossil fuel industry is arguably lengthier and more substantive than that of his likely rivals."[58] Alas, Walker ultimately dropped out of the race, while the Kochs' bête noire, Donald Trump, has secured the nomination of the Republican Party.

The Kochs have attempted to immunize themselves from criticism for their antienvironmental agenda and to curry favor with the science establishment through philanthropic contributions to science education and outreach-oriented institutions. They provide funding, for example, for the PBS science series *Nova*, produced by WGBH in Boston. That money appears to have earned David Koch a spot on the board of WGBH.

DAVID KOCH
"Climate does fluctuate. It goes from hot to cold. We have ice ages."

The Kochs also provide funding for the National Museum of Natural History in Washington, D.C., including $15 million from David Koch toward the cost of the new Hall of Human Origins. Some have argued that the exhibit seems to go out of its way to imply that climate change is beneficial rather than harmful.[59] Fifteen nongovernmental organizations recently launched a petition, signed by more than 100 leading scientists, calling on all museums of science and natural history to cut their ties with the Kochs and other funders of climate change denial.[60] Another organization has presented a petition with more than 100,000 signatures calling on WGBH to sever its ties with David Koch for similar reasons.[61]

As the Koch brothers have faced increased, unwanted scrutiny from the media over their antienvironmental agenda, among other things, their response seems to be "if you can't beat 'em, buy 'em"—or at least try. In 2013, they were reported to be planning to purchase the Tribune Company network of eight daily newspapers, which includes the widely circulated *Chicago Tribune* and the crown jewel of the Tribune family, the *Los Angeles Times*, with a Sunday circulation that exceeds 1 million readers. Clarence Page, a columnist for the *Chicago Tribune*, voiced the widely held fear among Tribune journalists that the Koch brothers would exploit the newspapers "as a vehicle for their political voice."[62] The takeover bid failed, however, after reports emerged that at least half of the *Los Angeles Times* staff would quit in the event of the takeover.[63]

"Don't Worry, Be Happy," or a Kinder, Gentler Denialism

The most insidious form of climate change denial, by some measure, is denial of the seriousness of the threat and monumental nature of the effort required if we are to avert dangerous climate change. As outright denial of the scientific evidence becomes ever less credible, a new breed of climate change denier, a kinder, gentler sort of denier, has appeared on the scene to exploit the new niche that is emerging in the world of climate change contrarianism.

The charismatic Bjorn Lomborg, the self-styled "skeptical environmentalist" who brandishes a Greenpeace T-shirt as evidence of his unassailable environmental bona fides, fills that niche to a T. In op-eds in leading newspapers, such as the *Wall Street Journal*, the *New York Times*, and *USA Today*, Lomborg professes his concern for the poor and his apprehension about the "energy poverty" they will suffer if we hinder continued efforts to extract and burn fossil fuels. He scolds those who would wean us off fossil fuels and incentivize the transition to a clean-energy economy.[64]

Lomborg's arguments often have a veneer of plausibility, but scratch the surface, and you witness a sleight of hand, where climate projections are lowballed; climate change impacts, damages, and costs are underestimated; and the huge current subsidies to the fossil fuel industry, both direct and indirect, are ignored.

Despite his professed sympathy for the plight of the developing world, Lomborg occasionally betrays his disregard for those most vulnerable to the impacts of climate change. In one op-ed, for example, he stated that "a 20-foot rise in sea levels . . . would inundate about 16,000 square miles of coastline, where more than 400 million people currently live. That's a lot of people, to be sure, but hardly all of mankind. In fact, it amounts to less than 6% of the world's population—which is to say that 94% of the population would not be inundated."[65] Yes—you read that right. What's 400 million people among friends?

And while Lomborg is expressing his deep concern for the poverty stricken of the world, he readily accepts a salary as high as $775,000 a year courtesy of a constellation of entities, including the Koch brothers, that fund his Copenhagen Consensus Center.[66]

"SKEPTICAL ENVIRONMENTALIST" BJORN LOMBORG
"On average, global warming is not going to harm the developing world."

The center is in fact a virtual entity, with an official address at a Lowell, Massachusetts, parcel service. The conservative government of Prime Minister Tony Abbott in Australia attempted to provide it with a permanent home, offering $4 million of taxpayer funds to the University of Western Australia if it would house the center. The university administration initially accepted the offer but rescinded the acceptance after widespread expression of outrage from faculty, who regarded the proposed center as undeserving of the institution's imprimatur.[67]

Lomborg is not alone. There are entire organizations in the United States, such as the so-called Breakthrough Institute, that seem entirely uninterested in any genuine "breakthroughs." They advocate for the continued exploitation of fossil fuels and are dismissive of the regulation of carbon emissions and of providing incentives for renewable energy. They argue that the free market will somewhat magically solve the problem on its own, without the need to factor in the externalities—the currently hidden (as far as the market is concerned) costs of burning fossil fuels.

Both Lomborg and the Breakthrough Institute seem very pessimistic about technological innovation and progress when it comes to widespread deployment of renewable energy, despite the remarkable progress that is already occurring with solar, wind, and geothermal energy. Yet they are extreme techno-optimists (or perhaps, more aptly, techno-Pollyannas) when it comes to the ostensible promise of as yet undemonstrated and untested schemes to manipulate the Earth system to offset greenhouse warming—what is known as *geoengineering*.[68]

Is geoengineering feasible? Would it be effective? Would it be safe? We address these questions in the next chapter.

The Judgment of History

History will judge the actors in this debate, and many will be judged harshly. By that time, unfortunately, it will be too late. The great gears of climate change, once set in motion, will inexorably be grinding toward the planet's demise. By then, the bad actors will have already accumulated their short-term personal gain, died, and passed it along to their children and grandchildren, who can only hope that everyone will have forgotten the role their ancestors played in the ensuing environmental degradation.

It is difficult to know whether climate change contrarians have taken their positions out of good faith, ignorance, willful ignorance, or calculated deceit. Whatever the reason, a host of individuals have used their public presence to confound and slow the response to climate change.

Even if the evidence were not as conclusive as it is, there is enough of it to give any educated person pause before dismissing the threat in its entirety or minimizing it to the point of public inaction or confusing it in the public mind to the point that the political process is stymied and the path forward blocked.

It is important to make note of who played what roles in this debate. If it turns out that, against all current evidence, climate change stops or reverses, the deniers can take their bows. If, however, they have helped lead us into disastrously misguided policy response, as it seems clear they have, history should not be allowed to forget who they are and what they have done.

I'VE ADDED A MIRROR, A PARTICLE CANNON, SOME IRON FILINGS AND A VACUUM. I THINK I'VE GOT IT RIGHT THIS TIME!
TOLES

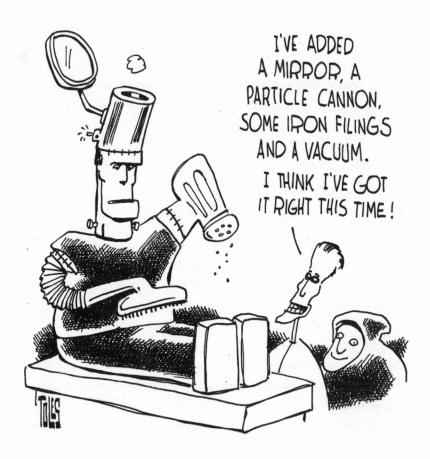

[7]

GEOENGINEERING, OR "WHAT COULD *POSSIBLY* GO WRONG?"

There was an old lady who swallowed a fly,
I don't know why she swallowed the fly,
Perhaps she'll die!

There was an old lady who swallowed a spider,
That wriggled and wiggled and tiggled inside her;
She swallowed the spider to catch the fly,
I don't know why she swallowed the fly,
Perhaps she'll die!

There was an old lady who swallowed a bird;
How absurd to swallow a bird.
She swallowed the bird to catch the spider,

. . .

There was an old lady who swallowed a cow,
I don't know how she swallowed a cow;
She swallowed the cow to catch the goat,
She swallowed the goat to catch the dog,
She swallowed the dog to catch the cat,

She swallowed the cat to catch the bird,
She swallowed the bird to catch the spider,
That wriggled and wiggled and tiggled inside her;
She swallowed the spider to catch the fly,
I don't know why she swallowed the fly,
Perhaps she'll die!

There was an old lady who swallowed a horse . . .
She's dead, of course!

Many of us grew up listening to and singing this song. As it turns out, the song is a perfect parable for the risks of *geoengineering*. In the old lady's case, the "solution" proves far more dangerous than the initial malady. Thus it potentially is with many of the technofix schemes proposed as purported solutions to the problem of human-caused climate change.

Searching for a Breakthrough

Many of those who advocate against taking action when it comes to dealing with the underlying problem—our ongoing burning of fossil fuels—have instead turned to possible technosolutions for counteracting climate change that involve other massive interventions in the Earth system: *geoengineering*. In some ways, for the free-market fundamentalist, geoengineering is a logical way out because it reflects an extension of faith that the free market and technological innovation can solve any problem we create, without the need for regulation.

Unsurprisingly, even many rather level-headed captains of industry, such as Bill Gates, have embraced the concept along with techno-Pollyannas, such as Bjorn Lomborg and the Breakthrough Institute.[1] Price on carbon? Nah, the market doesn't need it. Renewable energy? It's a pipe dream. Massively interfering with the Earth system in the

hope that we might get lucky and offset global warming? Yeah, that's the ticket!

It could very well be that we find ourselves in a situation where a stopgap measure is needed, where dangerous climate change is upon us and even worse impacts appear unavoidable, regardless of our best efforts to lower emissions. But every bit as likely—arguably more so—is that the prospect of geoengineering merely provides a crutch for critics of restraints on carbon emissions. Hey, climate, there's a simple remedy for what ails you, a sort of climate change methadone. No need to kick that carbon addiction after all. But as it turns out, that simple remedy ain't so simple. Seemingly the stuff of science fiction, many of the schemes that have been proposed—placing mirrors in space to reflect sunlight away from Earth, shooting reflective particles into the atmosphere to reduce the amount of incoming sunlight, "iron seeding" the oceans to get them to take up more CO_2 from the atmosphere, and quite literally sucking the CO_2 out of the atmosphere—come with potentially dangerous side effects. Such massive manipulations of the Earth system evoke the principle of unintended consequences. What, after all, could possibly go wrong? Well, a great deal, as it turns out. We could very well end up even worse off than if we hadn't engaged in these additional uncontrolled experiments with laboratory Earth.[2]

Mirrors . . . in . . . Space

One proposed geoengineering scheme involves placing a huge number of small mirrors in space to reflect some of the incoming sunlight back out to space—a planetary-scale "sunshade," if you will. If enough of these mirrors were placed above Earth's surface, they could reduce the incoming heating from the sun enough to offset the overall warming of Earth due to human-produced greenhouse gases.

If this idea sounds a bit like something from *Star Wars* to you, then you might not be surprised to learn that Cold War hawks such as Edward Teller, a major force behind President Ronald Reagan's proposed Strategic Defense Initiative ("Star Wars"), was a proponent of this geoengineering scheme.[3]

Sounds simple, right? Well, not really. The typical scheme involves placing many extremely thin individual mirrors, each less than 3 feet (1 meter) in diameter and weighing less than 0.035 ounce (1 gram), at a gravitationally stable location along the axis of Earth's orbit around the sun more than 620,000 miles (1 million kilometers) above Earth's surface. Trillions of these mirrors would be required to achieve the necessary cooling effect, and the logistics and expense of getting that many mirrors to a position that many miles above Earth are obviously prohibitive. By some estimates, it could cost as much as $350 trillion.[4] That amount is well over an order of magnitude larger than prevailing estimates of what it would cost to simply reduce our carbon emissions in the first place.

Even the primary advocate of this scheme, Roger Angel of the University of Arizona, while arguing that the approach may be needed as a last-ditch effort if we find ourselves barreling ahead toward a climate crisis, has conceded that "the sunshade is no substitute for developing renewable energy, the only permanent solution."[5]

Shooting Stuff into the Atmosphere

A related scheme involves shooting reflective particulates into the very stable upper part of the atmosphere, the stratosphere, where they can reside for several years. In principle, this process would mimic the way volcanic eruptions cool the planet. An explosive tropical volcanic eruption like that of Mount Pinatubo in 1991 puts enough reflective sulfate particulates into the stratosphere to cool the planet by about 1°F (0.6°C) for several years.

It turns out that it is quite feasible and not especially expensive to use large, specially built canons to shoot large amounts of these particulates into the stratosphere—as much as Pinatubo produced. If you do the calculations, you find that all it would take is a Pinatubo-size injection of particles every couple of years to offset the current warming effect of carbon emissions.

So, great, we've solved the feasibility and expense issues that proved so problematic with space mirrors. Problem solved, right? Not so quick.

It turns out that a number of problems emerge with this scheme.[6] One is pretty basic—if you implement this scheme, you don't get back the climate you started with. The spatial pattern of cooling due to a volcanic eruption—or, more to the point, due to a sulfate-injection scheme that mimics a volcanic eruption—isn't the mirror image of the pattern of greenhouse warming because the physics is different. In the case of a volcanic eruption, you are reducing the pattern of incident sunlight, whereas

in the case of greenhouse warming, you are preventing the escape of heat energy from Earth's surface. Those effects vary differently with respect to latitude and altitude.

The globe may not warm upon implementation of this scheme, but that's a consequence of a global averaging, wherein some regions will warm even faster than they were before the injection and some regions will actually cool. You read that right. *Some* regions will end up warming even faster. We could conceivably end up, for example, warming the southern oceans more rapidly, furthering the destabilization of the West Antarctic Ice Sheet and, with it, the acceleration of global sea-level rise. We can't rule out that scenario.

It turns out that the continents tend to cool relative to the oceans in model simulations. That pattern would reduce the vigor of the hydrological cycle over land. That's just a fancy way of saying that the continents would dry out. We might end up with worse droughts than if we had done nothing at all. So the question is: Do you feel lucky?

Meanwhile, those sulfate particles we're putting in the stratosphere have the potential to do some nasty things. As you may recall, it was the production of sulfate particles by industrial activity that created the acid rain problem in the 1960s and early 1970s before the passage of the Clean Air Act. The sulfate particles we are talking about in this case would be higher in the atmosphere (the stratosphere), above the region where raindrops form. But they ultimately would make it down into the lower atmosphere, into clouds and rainfall, and, finally, to the surface, where they would find their way into rivers and lakes.

OK, this solution would worsen the acid rain problem, but at least it won't worsen the other major global environmental problem of the past century, the hole in the ozone layer. Right? Wrong. It would worsen that problem, too. The sulfate particles would provide extra surface area for the ozone-depleting chemical reactions that take place in the stratosphere. Although the ozone layer has mostly recovered due to passage of the Montreal Protocol in the 1980s, there are still enough ozone-depleting Freon gases in the stratosphere that the extra kick they would get from the sulfates would likely continue the destruction of the protective layer.

And what about all that CO_2 that is continuing to accumulate in the atmosphere? Yes, we almost forgot about that. As with any "cover-up" solution to climate change that doesn't deal with the root cause of the problem, CO_2 would continue to build up not only in the atmosphere but in the oceans as well. The problem of ocean acidification—global warming's evil twin—would get continually worse. We would still have to say so long to the coral reefs.

That raises another issue. Assuming that we were to continue to burn fossil fuels, this "solution" would require us to continually shoot more and more sulfates into the stratosphere as CO_2 continues to accumulate in the atmosphere. It's a Faustian bargain if ever there was one. Most of that extra

CO_2 will remain in the atmosphere for thousands of years—CO_2 levels will be permanently elevated for all intents and purposes. What happens if there is a war, a plague, an asteroid collision, anything that might disrupt our technological infrastructure and interfere with our regular administering of sulfate injections? Within a matter of years, the reflective "cover" would disappear, and we would experience the full impact of decades' worth of greenhouse warming in a matter of years. That would give the term *abrupt climate change* a whole new meaning.

One other thing, by the way: reflecting more sunlight out to space before it ever reaches the surface of Earth means less potential for solar power, less availability of alternative energy. Geoengineering in this case would make even more difficult the already tough challenge of weaning ourselves off the fossil fuels that are at the very root of the problem we are trying to solve.

Dumping Stuff into the Ocean

If shooting stuff into the atmosphere turns out not to be the greatest idea, maybe dumping stuff into the ocean will work out better.

As it happens, one geoengineering scheme involves doing just that. The process is known as "iron fertilization" and is fairly straightforward. Over vast regions of the world's oceans, the primary limiting nutrient in the upper ocean is iron. If more iron were available, there would be more algae or "phytoplankton," which take up CO_2 when they photosynthesize. So by dumping modest amounts of iron dust into the ocean, we could potentially generate a *bloom* of phytoplankton activity, taking more CO_2 out of the atmosphere. As these phytoplankton continually die, they would, in principle, sink to the ocean bottom, where the carbon they gobbled up would be buried for the long term.

One of the seeming advantages of this approach is that it is solving the problem at its root cause, helping to take CO_2 out of the atmosphere. In principle, it also deals with the problem of ocean acidification. The idea seems so encouraging that a number of companies—Planktos and Climos being the most prominent—emerged on the scene in the past decade to commercialize it. Planktos was even so bold as to sell carbon credits—for $5, it would promise to take a ton of CO_2 out of the atmosphere, a seemingly cheap way for an individual, an organization, or a company to lower its effective carbon footprint.

Once again, though, the scheme hasn't quite delivered on its initial promise. Controlled field experiments have shown that iron fertilization at best leads to enhanced cycling of carbon in the upper ocean, with no apparent increase in deep carbon burial. Without deep burial, the removal of carbon from the atmosphere is only temporary. Even worse, some studies suggest that iron fertilization may actually favor harmful algae blooms, which are responsible for ocean dead zones and so-called red tides.

Without evidence of actual carbon burial, and with environmentalists and entire governments growingly increasingly concerned about rogue, uncontrolled, and potentially harmful iron-fertilization experiments being performed by companies such as Planktos, interest in the scheme has dried up.[7]

You might see a pattern emerging here.

A Giant Sucking Machine!

Let's continue with this theme. Although iron fertilization hasn't quite panned out, a scheme that works by removing carbon directly from the atmosphere is appealing for a number of reasons. Are there other ways to do it?

Certainly. Trees do it, after all.

Trees (and other plants) take carbon out of the atmosphere as they photosynthesize; store it in their trunks, branches, and leaves; and bury it in their roots, in the leaf and branch litter that falls and gets deposited on the forest floor, and in the soil. But the trees aren't very efficient at taking CO_2 out of the atmosphere. They, like us, respire CO_2. And when they die and decompose, they give some of their carbon back to the atmosphere.

What if we could make the "perfect" tree (from a climate standpoint, that is)? A *synthetic* tree that takes CO_2 out of the air using chemical processes that are 1,000 times more efficient than photosynthesis in their carbon-removal ability. A tree that doesn't give any of its carbon back to the atmosphere. Rather than dying and decomposing, it would turn the carbon into baking soda, which in principle can be buried for the long term. An array of 10 million of these trees distributed across the continents of the world could potentially serve as a "giant sucking machine," taking up a significant chunk (at least 10 percent) of our current carbon emissions.[8]

But . . . wait for it . . . there are some complications. After all, it's a lot harder to get the genie back in the bottle once you've let it out. In taking CO_2 out of the atmosphere, you're fighting the laws of thermodynamics, and that's a very expensive battle to wage. By some estimates, it would cost more than $500 per ton of carbon removed (though the cost could in principle be brought down substantially with additional research and through economies of scale). That's 100 times as much as what Planktos was charging for its scheme. The difference is that the artificial trees could actually work.

The prohibitive expense nonetheless means that at present it is *far easier and much less expensive to prevent the CO_2 from getting into the atmosphere in the first place,* whether by capturing and sequestering the carbon emitted from coal and natural gas power plants or, better yet, by getting our energy from renewable sources instead of from fossil fuels.

If, however, after doing everything possible to reduce our carbon emissions, we still find ourselves in need of a stop-gap scheme to avert catastrophic climate change, carbon-sucking artificial trees may be the safest and most efficacious of all the available geoengineering schemes out there.

What Could Possibly Go Wrong?

We have reviewed only some of the most prominent geoengineering schemes. There are other proposals, involving everything from painting rooftops white to seeding low clouds over the oceans. Certain things are common to nearly all of them.

With the possible exception of "direct air capture" (the technical term for the giant-sucking-machine scheme), each of the ideas comes with potential nasty surprises and the threat of unintended consequences. We could end up worse off than if we hadn't implemented these schemes at all.

The schemes are also fraught with political and ethical complications. For one thing, who gets to set the global thermostat? For low-lying island nations such as Tuvalu, current CO_2 levels are already too high—the island's inhabitants are threatened with the loss of their land and their rich cultural heritage by the several feet of sea-level rise that is already likely in train. While the industrial world debates whether we can still avoid "dangerous" warming of 3.6°F (2°C), dangerous warming is already here for many people on our planet. If they had their hands on the dial, they might want to set it at a lower temperature. Other nations might prefer it warmer. Who makes the decision?

One could easily imagine a whole new form of global conflict wherein rogue states employ geoengineering to control the climate for themselves. A climate model simulation might show, for example, that injecting sulfates into the stratosphere would relieve the drought that plagues a particular nation. Yet it would do so at the expense of causing a drought somewhere else. Perpetual conflict in the Middle East, it has been argued, has fundamentally always been about access to precious and scarce water resources.[9] Would geoengineering provide yet another weapon to be employed in this ongoing epic battle?

A Path We Want to Go Down?

The fundamental problem of geoengineering solutions is the monumental danger of tinkering with a complex system that we don't fully understand—Earth's climate system and the delicate, complex web of ecosystems that it supports. A crudely applied speculative mechanical fix might make things worse, not better. It is simply impossible to know or game out all the unintended consequences of deploying an untested technology on such a massive scale.

Compare this scenario with an experimental new treatment for a disease. "What's wrong with experimental medical treatments?" you ask. Well, consider the history of medicine. It is a history of many wrong turns and, unfortunately, many fatal results. Gains have accumulated over time, but a lot of time is required to get a treatment right. And in the case of global warming, there is only one patient, planet Earth, and we can't afford a fatality. There are no opportunities for controlled, randomized trials. There can be no control group. You may be treating the malady with aspirin, or you may be treating it with thalidomide. The proposed cure could well be worse than the disease. Indeed, it could prove fatal.

Although the threat the planet is facing is huge in scale, its cause is profoundly simple: an unhealthy dose of carbon dioxide. The simplest and safest solution is to address the problem at its root cause.

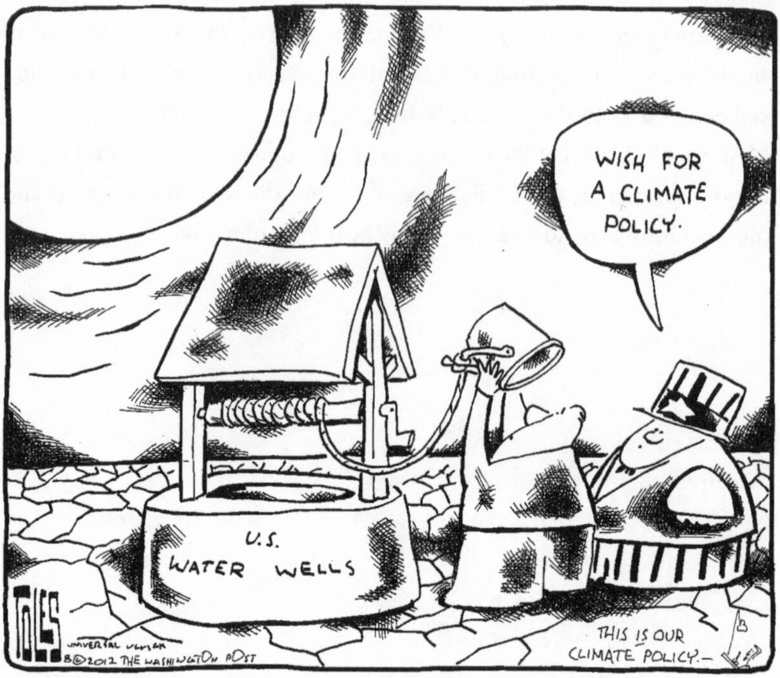
WISH FOR A CLIMATE POLICY.
U.S. WATER WELLS
THIS IS OUR CLIMATE POLICY.—
TOLES
UNIVERSAL
©2012 THE WASHINGTON POST

8·16·12

[8]

A PATH FORWARD

It would be easy to dismiss the challenge of averting dangerous climate change as insurmountable. It is a steep hill to climb if we are to avert what many consider to be the dangerous limit of 3.6°F (2°C) warming. Concerted action is required not a decade from now, not several years from now, but *now* if we are to have any hope of achieving that goal.

So there is indeed a great deal of urgency. The time for *wishing* for climate policy action has long passed. The time for *demanding* it has come.

The Black Double-Diamond Slope

Had we begun acting decades ago, we would have been in a position to slowly retreat from the burning of fossil fuels, proceeding gently down the bunny slope of decreasing carbon emissions. What decades of disinformation, denial, and delay have brought us to instead is a much steeper run down the black double-diamond slope. Our emissions have to come down rapidly.

Human beings currently emit more than 30 gigatons (30 billion tons) of CO_2 pollution a year. That number is difficult to fathom, so try this instead: Take all the elephants in the world. Although poaching has sadly led to a substantial decline in their numbers, about 500,000 of them are still left. Combining juveniles and adults, let's assume an average weight of 4 tons per elephant. That gives a total of 2 million tons. So the amount of CO_2 we emit per year is equivalent in weight to the total sum of the worlds' elephants multiplied by 15,000. You get it—we emit a lot of CO_2.

Given that there are around 7 billion people on Earth, the average person has a carbon footprint of about 4 tons (one elephant) a year. But that figure is somewhat misleading because there is a remarkably wide range between the most carbon-profligate and most carbon-frugal nations. Qataris have the highest average carbon footprint in the world at 40 tons (10 elephants) per person per year.[1] The United States comes in second, with an 18-ton (4.5-elephant) footprint per person, and China comes in third, with a 6-ton (1.5-elephant) footprint. Most developing nations come in under 1 ton (0.25 elephant).

Here's the challenge. If we want to avoid planetary warming of 3.6°F (2°C)—or what many observers consider "dangerous" warming, though, as we have noted, others might reasonably argue that's already too much—we have a very limited "carbon budget" left to work with. No more than 1 trillion or so tons of CO_2.

At the current rate of 30 gigatons a year, we'll burn through our budget in about three decades. To remain within the budget, we have to reduce emissions by several percent a year, bringing them down to 33 percent of current levels within twenty years. That's an average world-wide carbon footprint similar to what prevails in the developing world. By midcentury, emissions must approach zero. That's the black double-diamond slope.

One recent analysis determined that achieving these reductions would require that 33 percent of all proven reserves of oil, 50 percent of all natural gas, and 80 percent of all coal reserves must remain in the ground.[2] That means we have to phase out coal and leave most if not all of the Canadian tar sands in the ground (that is, no Keystone XL pipeline). And what about supposedly "clean" natural gas? Well, despite what is sometimes claimed, it appears quite clear that it is part of the problem, not the solution.

Executive Action

Monumental as the required reductions might seem, there is reason for cautious optimism that they can be achieved. Let's start with the developments here in the United States. It is true that House and Senate Republicans' intransigence has ensured that the passing of a comprehensive climate bill in the near future has few prospects. However, President Barack Obama has made skillful use of the executive branch to facilitate progress even in this hostile environment.

Let's look at the numbers. In the United States, roughly 33 percent of our carbon emissions come from the generation of electricity, and a little less than 33 percent comes from transportation. Add them up, and you have accounted for nearly 66 percent of our carbon emissions. The administration has targeted those emissions directly through executive actions.

The centerpiece of these efforts is the Environmental Protection Agency's new Clean Power Plan. Despite a concerted assault against it—in the form of lawsuits, attack ads, and the spread of misleading propaganda by lobbyists and oil-soaked politicians—the plan is now being implemented. Its goal is a 32 percent reduction in CO_2 emissions by power plants within roughly the next decade, but the plan provides some flexibility to individual states in how they go about achieving this reduction.[3]

States could require that coal power plants employ "carbon capture and sequestration" technology to remove the CO_2 exhaust before it makes its way into the atmosphere. The technology, however, is currently prohibitively expensive to deploy, rendering these practices uncompetitive in the energy marketplace.

Arguably far better would be the use of *biofuels* with carbon capture and burial, which leads to net removal of carbon from the atmosphere because the carbon that makes up the biofuels previously came *from* the atmosphere. Such an approach may fare well in states employing carbon-permit systems. But the required technology is not yet at the stage where it can be deployed on a wide scale.

Some have argued that switching to natural gas, which has a lower carbon footprint than coal per watt generated, might be part of the solution. But as we will see shortly, natural gas is not the "bridge" to a fossil fuel–free future that it is often advertised to be.

At present, the most viable solution is for states to decommission coal plants and introduce a larger share of renewables—solar, wind, and geothermal—into their energy portfolios to make up the difference. Additional strategies include reducing energy consumption and, finally, entering into regional carbon-trading consortiums, which we also discuss more fully later.

Critics have assailed the Clean Power Plan as a "war on coal." On his HBO show *Real Time with Bill Maher*, host Bill Maher justifiably asked one of us (Michael Mann), "What's wrong with a war on coal?" The reply: "We *do* have to move away from coal."[4] We need to shift away from our reliance on fossil fuels for power. The Clean Power Plan will help us do just that. Of course, at the same time we must help workers displaced by the phase-out of coal and other fossil fuels by supporting retraining programs to ensure that they find new opportunities in the growth industry of renewable energy and elsewhere in our economy.

Methane is an even more potent greenhouse gas than CO_2, more than thirty times as potent over the course of a century and nearly a hundred times worse on timescales of a couple of decades—a particular concern when it comes to the possibility of exceeding near-term climate tipping points. The primary source of methane in power generation is the fugitive emissions that leak out during the process of hydrologic fracturing (fracking) employed in extracting natural gas, which is mostly methane. Any comprehensive plan to decrease power-related greenhouse gas emissions has to target methane, too. Shortly after announcing the new Clean Power Plan, the EPA introduced another new set of measures aimed at lowering power-sector methane emissions by 45 percent by 2025.[5]

Equally bold were the efforts by the Obama administration during its first term to reduce transportation-related carbon emissions. In August 2012, the White House announced new fuel-efficiency standards requiring a fleet-averaged fuel efficiency of 55 miles per gallon for cars and light-duty trucks by 2025, which would represent a doubling of fuel efficiency relative to vehicles that were currently on the road and amount to a halving of gasoline consumption and, consequently, a halving of transport-related oil consumption.

Last but not least, the administration held the line on the Keystone XL pipeline. First, in February 2015, President Obama vetoed a congressional bill that attempted to greenlight the pipeline project, which would have ensured that extensive quantities of dirty, carbon-intensive tar sands oil from Canada would be mined and delivered to the global market over a period of decades.[6] Then, in November 2015, following a six-year-long State Department review, the president rejected the project, arguing that approval of the pipeline would "undercut" his administration's "global leadership" in "taking serious action to fight climate change."[7] It is possible that the Keystone project will be revived in the event of a future pro–fossil fuel administration. But for the time being, the effort is dead.

Grassroots Action

Coupled with efforts from the top down by the Obama administration to reduce carbon emissions, efforts by cities, municipalities, states, and regions from the bottom up have proved critical in the absence of congressional leadership on climate.

The mayors of three of the five largest cities in the United States—Los Angeles, Philadelphia, and Houston—joined together in September 2014 to form the Mayors' National Climate Action Agenda, which is aimed not only at achieving major reductions in carbon emissions in those cities but also at working collectively toward federal and international efforts to price carbon emissions. Since then, twenty-seven mayors have signed on, from San Francisco to Seattle to Kansas City to Austin to Columbus to Orlando. This coalition builds on a previous effort in 2007, the Mayors Climate Protection Agreement, which the mayors of more than 1,000 towns and cities around the United States signed.

There is perhaps no single town more symbolic of the grassroots, bottom-up approach to tackling climate change than Greensburg, Kansas, despite its modest population of fewer than 1,000 people. This aptly named town, though located in the reddest part of deep-red Kansas and led by a conservative Republican mayor, has become the model of a *green* American town. You couldn't ask for a more poetic story, right? But the story gets even better when you discover *how* this happened.

On the evening of May 4, 2007, Greensburg was devastated by an F5 tornado that leveled 95 percent of the town. Now let's leave aside, for the moment, the emerging evidence that climate change may be leading to more devastating tornadoes.[8] That's not the point of the story (though no doubt it provides an interesting subtext).

No, the story here is that Greensburg's Republican mayor, Bob Dixson, managed to turn this tragedy into a triumph, choosing to rebuild and reinvent Greensburg as a model of sustainability, now sometimes described as the "greenest city in America." The hospital, the city hall, and other town buildings were rebuilt to the highest certification level issued by Leadership in Energy and Environmental Design, meaning that they employ the most energy-efficient materials and technology available today. For his efforts, Dixson has been awarded the Mayor Richard M. Daley Legacy Award for Global Leadership in Creating Sustainable Cities.

Let's just pause for a moment to let that sink in. It's a feel-good story. A success story in the ongoing climate battle. And there are more of them, as we shall see.

Let's look at what is happening now at the state level. Despite subversive attempts by industry groups such as the American Legislative Exchange Council to sabotage efforts to cut carbon emissions and incentivize renewable energy, states are making significant progress.

Three West Coast states, which account for 15 percent of the U.S. population—California, Oregon, and Washington—have joined British Columbia in the Pacific Coast Action Plan on Climate and Energy to price carbon emissions through tradable-emissions permits. Thanks to the leadership of heroes such as California governor Jerry Brown, they are moving ahead whether Congress is willing to or not.

Nine northeastern states, which account for 13 percent of the U.S. population—Connecticut, Delaware, Maine, Maryland, Massachusetts, New Hampshire, New York, Rhode Island, and Vermont—have also banded together to form the Regional Greenhouse Gas Initiative (the number of states involved would have been ten, but, as readers may recall, Governor Chris Christie, in a concession to fossil fuel interests, pulled New Jersey out of the group). There is a good chance that other states—for example, Pennsylvania—may sign on.

In other words, states representing 28 percent of the U.S. population have already signed on to the pricing of carbon emissions. It is easy to imagine that number growing. With action now taking place at both the largest (national) and the smallest (city, state, and regional) scales, it will become irrelevant at some point whether Congress can get *its* act together or not.

So far we have talked only about the United States. Why not? It's a great country. We love this country. And we are rightfully proud of its historical heritage and the leadership in the world that it has displayed since its inception in 1776. What could be more tragic, though, than a nation of leaders becoming a nation of followers, trailing in the footsteps of other nations as they forge ahead? Yet that is precisely what we risk if we allow ourselves to fall behind the rest of the world in the race toward a green-energy economy.

We Americans have had the luxury of access to inexpensive energy from fossil fuels since our nation was founded. Oil, coal, and natural gas served us well during the Industrial Revolution, helping us to achieve our

modern energy-consumptive lifestyles. But the cost of their continued use is too great. They are a relic of generations past. They have no place in the modern age of high tech. It is time to move on.

Critics often claim that the real problem is rapidly industrializing countries such as China, which have moved ahead of us in total fossil fuel consumption (though they remain well behind us in *per capita* fossil fuel use). Why should we do anything, these critics ask, if they aren't?

Well, by many measures China is now doing far more than the United States to address the climate threat. It is spending considerably more on the research, development, and deployment of renewable energy. It has committed to a national carbon-pricing system.[9] As long as the Koch brothers and other fossil fuel interests maintain their stranglehold on the U.S. Congress, there is no hope for such a system here in the foreseeable future.

There is a larger point to be made, however. Given its legacy of two centuries of access to inexpensive, readily available fossil fuels, the United States has little credibility on the world stage in lecturing other nations on the importance of cutting carbon emissions and will have even less if it cannot get its own house in order. Who are we to tell the citizens of China, India, Brazil, and other developing nations with incipient energy economies that they don't deserve the same access to inexpensive, albeit dirty, energy sources if we do not display some leadership on this issue and make some concrete, productive changes in how we go about meeting our energy needs?

Fortunately, as we have seen, the United States is now displaying such leadership. The executive actions by the Obama administration coupled with the increasingly widespread action at the city, municipal, and state levels give the United States some credibility—additional geopolitical capital to expend when it sits at the table with other nations.

And expend that capital it has. In November 2014, a historic agreement was reached between the United States and China—the world's two largest emitters, which collectively account for nearly 50 percent of worldwide carbon emissions—to make substantial cuts in carbon emissions over the next two decades.[10] The United States has agreed to bring its emissions down by 26 to 28 percent (relative to 2005 levels) by 2025. China, having ramped up its fossil fuel economy only during the past decade or two, has committed to bringing its emissions to a peak by 2030. And for the doubters who might question China's resolve, it is worth noting that China is now reducing its coal use ahead of schedule.[11] This monumental commitment on the part of the world's two largest emitters of carbon sent an important message to other nations of the world going into the all-important United Nations (UN) Climate Change Conference in Paris in December 2015.

There is other good news. In 2014, for the first time in modern history, the global economy grew *without* any growth in carbon emissions, and in 2015 we saw an actual *drop* in global carbon emissions while the economy continued to grow—a powerful rebuttal to those who argue that it is impossible to decouple the global economy from fossil fuels.[12]

Although pundits will continue to debate the precise factors behind this trend, there is little doubt that the transition toward renewable energy already under way has played an important role here. Renewables are currently adding greater capacity than fossil fuels. They have now exceeded fossil fuel energy in added capacity worldwide for several years and are providing nearly 25 percent of all electricity. In some countries, such as Germany, the figure is closer to 33 percent. In the United States, the portion is lower, just less than 15 percent,[13] but we are seeing dramatically increased adoption of renewable energy.

Solar power more than doubled in the United States from 2013 to 2014,[14] and wind energy increased by just less than 10 percent.[15] Both solar and wind are on a trajectory to achieve "grid parity" in less than a decade—that's just a fancy way of saying that they will be competitive with fossil fuel energy even *without* a price on carbon.[16] States such as California, where solar is huge, and Texas, which is big on wind, have already achieved parity. So will the rest of the country if we level the playing field by putting a price on carbon that reflects the damage done by the burning of fossil fuels.

In the United States, we are also seeing remarkable improvements in energy efficiency[17] and a huge spike in sales of electric vehicles and hybrids, such as the Toyota Prius, the Tesla Model S, and the Nissan Leaf.[18] Despite the fierce pushback from fossil fuel interests and those doing their bidding, the progress has been palpable. A shift in the political winds seems to be in the air.

In the wake of the Climate Change Conference in Paris, there was a collective feeling of euphoria throughout the world that maybe, just maybe, we're now finally ready to turn the corner in confronting the climate challenge. The summit produced for the first time a deal that would aim to keep warming lower than the dangerous 3.6°F (2°C) limit, with an aspirational goal of an even lower limit (2.7°F [1.5°C]), in recognition of the threat of near-term global sea-level rise. For the first time, there was unanimous buy-in from all (197) participating nations to lower their carbon emissions in the years ahead, including industrial nations and developing countries alike. Even oil-rich Saudi Arabia climbed on board. The emissions reductions agreed on in Paris will not on their own stabilize warming lower than the dangerous 3.6°F limit, but the agreement puts in place a framework for negotiations on more stringent reductions at subsequent conferences. A path to averting catastrophic warming of the planet now seems possible.[19]

Politics of Change

We have seen that progress has already been made, here in the United States and throughout the world, in moving away from the fossil fuel–driven economy that is putting our climate and our future at risk. We are seeing action at the international level and in the United States at the presidential level, the state level, the local level, and, just as important, the individual level.

On September 21, 2014, in advance of the UN Climate Change Conference in New York, more than 300,000 people marched through the streets of that city in the largest demonstration on climate change in history. Similar marches took place in cities around the world in what has come to be known as the "People's Climate March."[20] Its purpose was to raise awareness and demand action from policy makers in averting a climate crisis.

Among those who marched were UN secretary-general Ban Ki-moon, former vice president Al Gore, New York City mayor Bill de Blasio, and a number of celebrities, who joined thousands of others from around the country and the world. To open the summit, Leonardo DiCaprio, who had been named the UN Messenger of Peace for his commitment to the climate issue, gave a powerful and inspiring speech about the urgency of action.[21]

In the wake of the climate march, the Rockefeller Brothers Fund, representing the estate of John D. Rockefeller—who founded the oil empire that would eventually become ExxonMobil—announced that it would be divesting itself of all fossil fuel holdings. The symbolic significance of this act was undeniable, and it was the culmination of an ongoing divestment campaign that now amounts to more than $50 billion in funds withdrawn by more than 300 foundations, faith-based institutions, government organizations, colleges, and financial institutions around the world.[22]

The divestment campaign is the brainchild of climate activist Bill McKibben, who founded the organization 350.org, which derives its name from the level of CO_2 in the air, 350 ppm, that climate scientist James Hansen has argued is the safe limit (a level that is substantially lower than the *current* level, let alone the levels we're headed toward in the absence of climate policy).[23] Although divestment may in the end be more important for its symbolic value than for any actual financial impact it will have on the fossil fuel industry, it has a very practical justification.

By some estimates, the fossil fuel industry currently has five times as much oil, gas, and coal in proven reserves than we can afford to burn if we are to avoid 3.6°F (2°C) warming of the planet.[24] If one accepts the premise that we cannot possibly afford to burn all those fossil fuels, then these companies have a fatal liability—most of their key assets must ultimately be stranded. They are a bad investment not only for the planet but also for the investor. That isn't even taking into account other emerging liabilities. ExxonMobil, for example, has recently found itself under investigation for knowingly hiding the adverse impacts of its product, fossil fuels,[25] much as the tobacco industry was investigated decades ago for hiding the harmful effects of its product. Cigarette makers ended up agreeing to a $370 billion settlement in a lawsuit brought by a group of state attorneys general in the late 1990s.[26] The attorney general of New York is now looking into the possibility of bringing a similar lawsuit against ExxonMobil.[27]

Folks such as McKibben and Gore—who, although having retreated somewhat from the public spotlight, continues to raise climate awareness through his Climate Reality venture and other activities—have played an important role in galvanizing the climate change troops. Their efforts, however, have served mostly to motivate those who already care about the issue.

But other organizations, institutions, and people—including the U.S. military, the insurance and reinsurance industries, and prominent members of the business community who recognize the serious threat posed by climate change and have increasingly spoken out about it—have helped to expand the base by raising awareness among those who might not otherwise be predisposed to care much about something they see as a purely "environmental issue."

Along with Pope Francis's recent campaign to raise awareness among the 77 million self-identified Catholics in the United States (and the larger audience of Americans who are listening to his message), we may just be seeing the formation of a coalition for climate action in the United States. This domestic coalition feeds into an even broader international one due

to the Paris Climate Change Conference, wherein the nations of the world have now joined together in common cause to wean civilization off fossil fuel energy in time to avert the disastrous warming of the planet.

The political challenge in the United States nonetheless remains daunting. Fossil fuel interests, represented by the notorious Charles and David Koch, have done their best to purge climate moderates from the Republican Party, and outright climate change denial remains stronger than ever in Congress. Yet a number of very prominent Republicans, albeit no longer in office, have publicly campaigned for a sober, fact-based discussion about climate change.

There is Sherwood Boehlert (R-N.Y.), former chair of the House Committee on Science, Space, and Technology. Facing an opponent funded by polluting interests, he retired from Congress in 2007. A champion of environmental causes, he has remained active on climate by participating on the board of the bipartisan organization Alliance for Climate Protection and by penning op-eds such as the one he wrote for the *Washington Post* in 2010—"Can the Party of Reagan Accept the Science of Climate Change?"—urging moderation on the part of his fellow Republicans when it comes to climate change.[28]

Former representative Bob Inglis (R-S.C.) had a near-perfect conservative voting record while in Congress (the American Conservative Union gave him a lifetime voting score of 93.5 percent). But he made the mistake of speaking out, as an evangelical Christian concerned about our stewardship of Earth, regarding the importance of acting on climate change. For this sin, he was "primaried" out of a safe congressional seat by a Koch brothers–supported "Tea Party Republican," Trey Gowdy. Inglis went on to found and direct the Energy and Enterprise Initiative, aimed at convincing conservatives to accept the science of climate change and at promoting market-based solutions to the problem. He recently received the John F. Kennedy Profile in Courage Award for his efforts.[29]

As you may recall from an earlier chapter, Republican presidents Richard Nixon, Ronald Reagan, and George H. W. Bush all supported regulatory solutions to environmental threats. In 2013, the four EPA administrators who served under these presidents wrote an op-ed in the *New York Times*: "A Republican Case for Climate Action."[30] In a *Washington Post* op-ed in 2015, "A Reagan Approach to Climate Change," George Shultz, secretary of state in the Reagan administration, argued that Reagan would have acted on climate change.[31]

Republican Arnold Schwarzenegger, who made climate action a focal point of his governorship of California, now coproduces the Showtime series *Years of Living Dangerously*, which highlights the threats from climate change.

Still in the Senate are John McCain (R-Ariz.) and Lindsey Graham (R-S.C.), who are firmly on record as supporting action on climate change. But they have been relatively quiet about the issue in recent years, presumably aware of the toxic nature of the issue to the interests that currently control the Republican Party.

Can all these climate moderates reclaim their party and restore reason when it comes to discussions of climate change within the Republican Party? They may have to.

The next election is arguably a make-or-break election for the climate. If we are to avert the worst impacts of climate change, we will have to act in the next few years. We will need both the president and Congress on board if we're going to work with other nations of the world to achieve the sorts of carbon reductions necessary to achieve this goal.

What Can I Do?

The problem of climate change is so large and contentious that it is tempting to avoid thinking about it at all. But avoidance is no longer an option. Once you do force yourself to think about it, where should you begin?

Leave the madhouse. Stop equivocating when discussing the science. If you meet someone who says that there is no warming or that the facts are not known, don't argue the point. Simply say politely that denial is no longer a respectable position because it's not. If he asks for evidence, hand him this book. Direct her to the reports by the Intergovernmental Panel on

Climate Change and the National Academy of Sciences. If he says that all science is suspect, tell him that view smacks of paranoid conspiratorialism, and don't debate it any further. Move along to a sensible person who may actually want to help solve the problem.

The science on climate is solid. It has been studied from every angle and over a long period of time. It is not merely a matter of climate scientists having reached a consensus. The overwhelming preponderance of facts is *in consensus*. These facts can be obscured for only so long. People are now seeing the impacts of climate change with their own two eyes. The tide is now turning. As with tobacco or air pollution, it's time to move on to talking about the solutions.

Support renewable energy and a price on carbon, and vote for representatives who will do the same. Pricing carbon isn't some unfair, arbitrary, punitive tax. It merely puts a legitimate price on a seriously dangerous waste product currently being dumped free of charge into our common atmosphere. It's no different from putting a price on collecting garbage rather than letting people dump theirs onto your street. Pricing carbon is a market-driven approach to solving the problem, so political conservatives ought to be able to support it.

Renewables are crossing into real viability even as you read this book. But the more support they get, the more rightful subsidies they receive, and the more use they get, the quicker the economies of scale will kick in and position them to take on the task of meeting an increasing societal demand for energy. If you encounter a cynic who insists that renewables aren't up to the task, you might quote the father of fossil fuel–power generation, Thomas Edison, who, giving new meaning to the word *prescience*, once said, "I'd put my money on the sun and solar energy. What a source of power! I hope we don't have to wait until oil and coal run out before we tackle that."[32]

If Edison isn't your cup of tea, too retro perhaps for your technomodernist sensibilities, we give you instead the great Elon Musk, entrepreneur, inventor, and engineer of SpaceX, PayPal, and Tesla fame. In explaining the rationale behind his latest project, the solar-power venture Solar City, he put it this way: "The sun, that highly convenient

and free fusion reactor in the sky, radiates more energy to the Earth in a few hours than the entire human population consumes from all sources in a year. This means that solar panels, paired with batteries to enable power at night, can produce several orders of magnitude more electricity than is consumed by the entirety of human civilization."[33] It's sort of hard to argue with that kind of logic.

Join an organization with a good track record on climate. The Union of Concerned Scientists comes to the mind, for example, but there are many others. Membership in such organizations will help ensure that you stay informed. It will also amplify your voice in the public debate and support an agenda of action.

Vote climate! We have to elect politicians who will vote in our interests rather than for the special interests. Alas, in the current presidential contest, we could not have a more stark choice before us, between a candidate who rejects the overwhelming evidence that climate change is happening and a candidate who embraces the role of a price on carbon and incentives for renewable energy. If you care about the planet, the choice would seem clear.

Help facilitate the next stage of evolution in the battle for environmental sustainability. The first stages were about toxins in our rivers and poisons in our air. They were about recycling and eating well and sustainably. This was an environmentalism about the health of local systems. But we now need something bigger.

Now we need a *whole-planet* environmentalism because the threat is to every living thing on Earth. The era of space exploration has taught us many things. It is time to pay attention to what we have learned. We have learned how extraordinarily rare a planet like Earth is and what a narrow, delicate

range of conditions sustain life. Although we can dream of someday finding other worlds like ours, we know now that they are unfathomably far away. The future of humanity is right here on Earth.

The most misguided notion spawned by the era of space exploration is—you've no doubt heard it before—"We are going to wreck this planet, so we need to find a new one." No. The population of Earth will *not* be moving away to another planet—not in your lifetime, not in your children's lifetime, and not in your great-great-great-great-grandchildren's lifetime. Nor will Earth's other irreplaceable species be making the move.

We will not, we *cannot*, wreck this planet. There is no Planet B. Earth is a rarity of literally cosmic proportions. It is an overflowing treasure chest of life-forms of unimaginable variety and beauty. It is perfectly fitted to us as humans because we evolved to fit it. It would amount to the gravest criminal act of irresponsibility in human history were we to throw it into fatal imbalance because of a wanton addiction to carbon.

So we have our work before us. We have our task. We have the oceans to preserve. We have the rain forests to protect. We have farmlands and coasts to defend. We have the panoply of spectacular species with which we evolved to shepherd. *This is our home. It's time to start acting like it.*

NOTES

1. Science

1. Carl Sagan, *Cosmos* (New York: Random House, 1980), 333.
2. Carl Sagan, writer and host, "Encyclopaedia Galactica," Episode 12 of *Cosmos: A Personal Voyage*, PBS, December 14, 1980.
3. Carl Sagan, *Broca's Brain: Reflections on the Romance of Science* (New York: Random House, 1979), 64.
4. See, for example, Michael Mann, *The Hockey Stick and the Climate Wars: Dispatches from the Front Lines* (New York: Columbia University Press, 2013), specifically 202–204 on industry-funded critic Paul Driessen.
5. The episode was summarized in Malcolm W. Browne, "Physicists Debunk Claim of a New Kind of Fusion," *New York Times*, May 3, 1989.
6. For a detailed discussion, see Mann, *Hockey Stick and the Climate Wars*.
7. Both the original hockey stick study (M. E. Mann, R. S. Bradley, and M. K. Hughes, "Northern Hemisphere Temperatures During the Past Millennium: Inferences, Uncertainties, and Limitations," *Geophysical Research Letters* 26 [1999]: 759–762), and the third report of the Intergovernmental Panel on Climate Change (IPCC) (*Climate Change 2001: The Scientific Basis. Contribution of Working Group I to the Third Assessment Report of the Intergovernmental Panel on Climate Change*, ed. J. J. McCarthy et al. [Cambridge: Cambridge University Press, 2001]) concluded that recent average warmth in the Northern Hemisphere was *likely* unprecedented for the *past 1,000 years*. The IPCC's fourth report (*Climate Change 2007: The Physical Science Basis. Contribution of*

Working Group I to the Fourth Assessment Report of the Intergovernmental Panel on Climate Change, ed. S. Solomon et al. [Cambridge: Cambridge University Press, 2007]) extended that conclusion back further, over the *past 1,300 years,* and raised the confidence to "very likely" for the past 400 years based on the availability of more than a dozen studies confirming that conclusion. The IPCC's most recent report (*Climate Change 2013: The Physical Science Basis. Contribution of Working Group I to the Fifth Assessment Report of the Intergovernmental Panel on Climate Change,* ed. T. F. Stocker et al. [Cambridge: Cambridge University Press, 2013]) extended the conclusion back over the *past 1,400 years.* By any honest reading, the IPCC's assessment, which represents the most thorough and up-to-date judgment of climate science, has not only reaffirmed but substantially strengthened and extended the original hockey stick conclusions.

8. In 2012, a team of seventy-eight leading paleoclimate scientists representing the PAGES 2k consortium, drawing on the most widespread paleoclimate database that has been assembled, published a new reconstruction of large-scale temperature trends (PAGES 2k Network, "Continental-Scale Temperature Variability During the Past Two Millennia," *Nature Geoscience 6* [2013]: 339–346, doi:10.1038/NGEO1797). They concluded that the recent global warmth is unprecedented in at least 1,300 years. A direct comparison by German paleoclimatologist Stefan Rahmstorf reveals these scientists' temperature reconstruction to be virtually identical to the original hockey stick reconstruction ("Most Comprehensive Paleoclimate Reconstruction Confirms Hockey Stick," *Climate Progress,* July 8, 2013, http://thinkprogress.org/climate/2013/07/08/2261531 /most-comprehensive-paleoclimate-reconstruction-confirms-hockey-stick/).

9. For an excellent summary of the tobacco industry's efforts, see Naomi Oreskes and Eric M. Conway, *Merchants of Doubt: How a Handful of Scientists Obscured the Truth on Issues from Tobacco Smoke to Global Warming* (New York: Bloomsbury Press, 2010), chap. 1, "Doubt Is Our Product."

10. Mann, *Hockey Stick and the Climate Wars.*

11. Chris Mooney, *The Republican War on Science* (New York: Basic Books, 2005).

12. National Academy of Sciences, *Climate Change Science: An Analysis of Some Key Questions* (Washington, D.C.: National Academy of Sciences, 2001); Royal Society, "Joint Science Academies' Statement: Global Response to Climate Change," June 7, 2005, https://royalsociety.org/topics-policy /publications/2005/global-response-climate-change/.

13. A list of the various scientific societies and academies that have endorsed the basic assessment on human-caused climate change is available from the Union

of Concerned Scientists, "Scientific Consensus on Global Warming," http://
www.ucsusa.org/global_warming/science_and_impacts/science/scientific
-consensus-on.html.

14. David Michaels, *Doubt Is Their Product: How Industry's Assault on Science Threatens Your Health* (New York: Oxford University Press, 2008), 11.

15. John Oliver, "97% to 3% Climate Change Debate," *Last Week Tonight*, HBO, May 11, 2014, https://www.youtube.com/watch?v=cjuGCJJUGsg.

16. John Cook et al., "Quantifying the Consensus on Anthropogenic Global Warming in the Scientific Literature," *Environmental Research Letters* 8 (2013), doi:10.1088/1748-9326/8/2/024024.

17. See, for example, Joe Romm, "Faux Pause: Ocean Warming, Sea Level Rise, and Polar Ice Melt Speed Up, Surface Warming to Follow," *Climate Progress*, September 25, 2013, http://thinkprogress.org/climate/2013/09/25/2562441/faux-pause-ocean-warming-speed-up/.

2. Climate Change

1. Spencer R. Weart, *The Discovery of Global Warming*, rev. ed. (Cambridge, Mass.: Harvard University Press, 2008), 7.

2. Sources for the scientific evidence discussed in this chapter include Michael E. Mann and Lee R. Kump, *Dire Predictions: Understanding Climate Change*, 2nd ed. (New York: DK, 2015); and Intergovernmental Panel on Climate Change, "Summary for Policymakers," in *Climate Change 2013: The Physical Science Basis. Contribution of Working Group I to the Fifth Assessment Report of the Intergovernmental Panel on Climate Change*, ed. T. F. Stocker et al. (Cambridge: Cambridge University Press, 2013), 1–27.

3. Michael E. Mann and Peter H. Gleick, "Climate Change and California Drought in the 21st Century," *Proceedings of the National Academy of Sciences* 112 (2015): 3858–3859.

4. J. O. Sewall and L. C. Sloan, "Disappearing Arctic Sea Ice Reduces Available Water in the American West," *Geophysical Research Letters* 31 (2004), doi:10.1029/2003GL019133.

5. C. H. Greene, J. A. Francis, and B. C. Monger, "Superstorm Sandy: A Series of Unfortunate Events?" *Oceanography* 26 (2013): 8–9.

6. Stefan Rahmstorf et al., "Evidence for an Exceptional 20th-Century Slowdown in Atlantic Ocean Overturning," *Nature Climate Change* 5 (2015): 475–480.

7. N. S. Diffenbaugh, M. Scherer, and R. J. Trapp, "Robust Increases in Severe Thunderstorm Environments in Response to Greenhouse Forcing,"

Proceedings of the National Academy of Sciences 110 (2013): 16361–16366; J. B. Elsner, S. C. Elsner, and T. H. Jagger, "The Increasing Efficiency of Tornado Days in the United States," *Climate Dynamics* 45 (2015): 651–659.

8. Michael Mann, *The Hockey Stick and the Climate Wars: Dispatches from the Front Lines* (New York: Columbia University Press, 2013), 84–87.

9. See, for example, K. Emanuel, "Downscaling CMIP5 Climate Models Shows Increased Tropical Cyclone Activity over the 21st Century," *Proceedings of the National Academy of Sciences* 110 (2013): 12219–12224, doi:10.1073/pnas .1301293110.

10. See, for example, K. E. Trenberth, J. T. Fasullo, and T. G. Shepherd, "Attribution of Climate Extreme Events" [review], *Nature Climate Change* 5 (2015): 725–730, doi:10.1038/NCLIMATE2657.

11. R. Horton et al., "New York City Panel on Climate Change 2015 Report: Sea Level Rise and Coastal Storms," *Annals of the New York Academy of Sciences* 1350 (2015): 36–44, doi:10.1111/nyas.12593.

12. H. Leifert, "Sea Level Rise Added $2 Billion to Sandy's Toll in New York City," *Eos* 96 (2015), doi:10.1029/2015EO026349.

13. Michael E. Mann, "False Hope," *Scientific American*, April 2014, 78–81.

14. E. Rignot et al., "Widespread, Rapid Grounding Line Retreat of Pine Island, Thwaites, Smith, and Kohler Glaciers, West Antarctica, from 1992 to 2011," *Geophysical Research Letters* 41 (2014): 3502–3509, doi:10.1002/2014GL060140.

15. James Hansen et al., "Assessing 'Dangerous Climate Change': Required Reduction of Carbon Emissions to Protect Young People, Future Generations, and Nature," *PLoS ONE* 8, doi:10.1371/journal.pone.0081648; A. Dutton et al., "Sea-Level Rise Due to Polar Ice-Sheet Mass Loss During Past Warm Periods," *Science* 349 (2013): 153–162, doi:10.1126/science.aaa4019.

3. Why Should I Give a Damn?

1. Detailed discussions of the impacts of climate change are provided in Michael E. Mann and Lee R. Kump, *Dire Predictions: Understanding Climate Change*, 2nd ed. (New York: DK, 2015); and Intergovernmental Panel on Climate Change (IPCC), "Summary for Policymakers," in *Climate Change 2014: Impacts, Adaptation, and Vulnerability. Part A: Global and Sectoral Aspects. Contribution of Working Group II to the Fifth Assessment Report of the Intergovernmental Panel on Climate Change*, ed. C. B. Field et al. (Cambridge: Cambridge University Press, 2014), 1–32.

2. See, for example, Peter Gleick, "Water, Drought, Climate Change, and Conflict in Syria," *Weather, Climate & Society* 6 (2014): 331–340.

3. Brad Plumer, "New Forecast: The Earth Could Have 11 Billion People by 2100," *Vox*, September 18, 2014.

4. World Food Programme, "Hunger," 2016, https://www.wfp.org/hunger.

5. John Vidal, "Millions Face Starvation as World Warms, Say Scientists," *Guardian*, April 13, 2013, http://www.theguardian.com/global-development/2013/apr/13/climate-change-millions-starvation-scientists.

6. See, for example, "In Hot Water: Columbia's Sockeye Salmon Face Mass Die-off," *Al Jazeera America*, July 27, 2015, http://america.aljazeera.com/articles/2015/7/27/half-of-columbia-rivers-sockeye-salmon-dying-due-to-heat.html.

7. Craig Welch, "Seachange: Oysters Dying as Coast Hit Hard," *Seattle Times*, September 11, 2013.

8. "Ogallala Aquifer," BBC News (online), n.d., http://news.bbc.co.uk/2/shared/spl/hi/world/03/world_forum/water/html/ogallala_aquifer.stm.

9. See, for example, R. Singh et al., "A Vulnerability Driven Approach to Identify Adverse Climate and Land Use Change Combinations for Critical Hydrologic Indicator Thresholds—Application to a Watershed in Pennsylvania, USA," *Water Resources Research* 50 (2014): 3409–3427, doi:10.1002/2013WR014988.

10. Jeff Masters, "U.S. Storm Surge Records," Weather Underground, n.d., http://www.wunderground.com/hurricane/surge_us_records.asp.

11. Such is the conclusion given in N. Lin et al., "Physically Based Assessment of Hurricane Surge Threat Under Climate Change," *Nature Climate Change* 2 (2012): 462–467.

12. Emma Batha, "Sahel Region Set to See Rise in 'Climate Refugees,'" Reuters, August 2, 2013, http://news.trust.org//item/20130802101500-bklf3/.

13. White House, *White House Report: The National Security Implications of a Changing Climate*, May 20, 2015, https://www.whitehouse.gov/the-press-office/2015/05/20/white-house-report-national-security-implications-changing-climate.

14. Much of the data in this section can be found in DARA and the Climate Vulnerable Forum, *Climate Vulnerability Monitor: A Guide to the Cold Calculus of a Hot Planet*, 2nd ed. (Madrid: Fundación DARA Internacional, 2012).

15. K. E. Trenberth, J. T. Fasullo, and T. G. Shepherd, "Attribution of Climate Extreme Events," *Nature Climate Change* 5 (2015): 725–730, http://dx.doi.org/10.1038/nclimate2657.

16. Brandon Keim, "Russian Heat, Asian Floods May Be Linked," *Wired*, August 10, 2010, http://www.wired.com/2010/08/russian-heat-asian-floods/.

17. Vladimir Petoukhov et al., "Quasiresonant Amplification of Planetary Waves and Recent Northern Hemisphere Weather Extremes," *Proceedings of the National Academy of Sciences* 110 (2013): 5336–5341. See also Stephen Leahy, "Killer Heat Waves and Floods Linked to Climate Change," Inter Press Service, February 27, 2013, http://www.ipsnews.net/2013/02/killer-heat -waves-and-floods-linked-to-climate-change/.

18. Quoted in "NASA Chief Questions Urgency of Global Warming," *Morning Edition*, National Public Radio, May 31, 2007, http://www.npr.org/templates /story/story.php?storyId=10571499.

19. National Climatic Data Center, National Oceanic and Atmospheric Adminis-tration, "Billion-Dollar Weather/Climate Disasters," n.d., https://www.ncdc .noaa.gov/billions/summary-stats.

20. H. Leifert, "Sea Level Rise Added $2 Billion to Sandy's Toll in New York City," *Eos* 96 (2015), doi:10.1029/2015EO026349.

21. Bob Berwyn, "Ski Industry Sees $1 billion in Global Warming Losses," *Summit Voice*, December 6, 2012, http://summitcountyvoice.com/2012/12/06/report -ski-industry-sees-1-billion-in-global-warming-losses/.

22. Wendy Koch, "Climate Change, Extreme Weather Spike Food Prices," *USA Today*, November 28, 2011, http://content.usatoday.com/communities /greenhouse/post/2011/11/climate-change-extreme-weather-spike-food -prices/1#.Va_Ef7eGpK4.

23. DARA and the Climate Vulnerable Forum, *Climate Vulnerability Monitor*.

24. IPCC, "Summary for Policymakers."

25. M. L. Weitzman, "Fat Tails and the Social Cost of Carbon," *American Economic Review* 104 (2014): 544–546.

26. I. Allison et al., *The Copenhagen Diagnosis, 2009: Updating the World on the Latest Climate Science* (Sydney: Climate Change Research Centre, University of New South Wales, 2009).

27. Pope Francis, "Address of His Holiness Pope Francis to the Members of the Diplomatic Corps Accredited to the Holy See," January 13, 2014, English trans-lation provided by the Vatican at http://w2.vatican.va/content/francesco /en/speeches/2014/january/documents/papa-francesco_20140113_corpo -diplomatico.html.

28. Quoted in Kimberly Winston, "Pope Francis Fan Club Takes to the Media After Encyclical," Religion News Service, June 18, 2015.

29. Marcia McNutt, "The Beyond-Two-Degree Inferno" [editorial], *Science*, July 3, 2015, 7, doi:10.1126/science.aac8698.

4. The Stages of Denial

1. Consider the case of a German high-school teacher who published an article, in a contrarian journal, based on a small number of nonrepresentative and contaminated urban records to argue against the rise in atmospheric CO_2 levels (Ernst-Georg Beck, "180 Years of Atmospheric CO_2 Gas Analysis by Chemical Methods," *Energy and Environment* 18 [2007]: 259–282). For a more detailed discussion of this case, see Michael Mann, *The Hockey Stick and the Climate Wars: Dispatches from the Front Lines* (New York: Columbia University Press, 2013), 25.
2. The matter is discussed in detail in Mann, *Hockey Stick and the Climate Wars*, 181–183.
3. The origins of this dubious approach appear to be an e-mail from retired MIT climate change contrarian Richard Lindzen leaked by climate change denier Anthony Watts on his blog: "Look at the attached. There has been no warming since 1997 and no statistically significant warming since 1995. Why bother with the arguments about an El Nino [*sic*] anomaly in 1998? . . . Best wishes, Dick" ("A Note from Richard Lindzen on Statistically Significant Warming," *WUWT: Watts Up With That*, March 11, 2008, http://wattsupwiththat.com/2008/03/11/a-note-from-richard-lindzen-on-statistically-significant-warming).
4. Stephan Lewandowsky et al., "Seepage: Climate Change Denial and Its Effect on the Scientific Community," *Global Environmental Change* 33 (2015): 1–13.
5. Michael E. Mann, "False Hope: Earth Will Cross the Climate Danger Threshold by 2036," *Scientific American*, March 18, 2014, http://www.scientificamerican.com/article/earth-will-cross-the-climate-danger-threshold-by-2036/.
6. See, for example, Byron A. Steinman, Michael E. Mann, and Sonya K. Miller, "Atlantic and Pacific Multidecadal Oscillations and Northern Hemisphere Temperatures," *Science* 347 (2015): 988–991. For a nontechnical discussion of this paper, see Michael E. Mann, "Climate Oscillations and the Global Warming Faux Pause," *Huffington Post*, February 26, 2015, http://www.huffingtonpost.com/michael-e-mann/climate-change-pause_b_6671076.html.
7. Mann, *Hockey Stick and the Climate Wars*, 57.

8. In a profile, journalist Fred Guterl noted that Lindzen "clearly relishes the role of naysayer. He'll even expound on how weakly lung cancer is linked to cigarette smoking" ("The Truth About Global Warming: The Forecasts of Doom Are Mostly Guesswork, Richard Lindzen Argues—and He Has Bush's Ear," *Newsweek*, July 23, 2001). Former NASA chief climate scientist James Hansen recalls meeting Lindzen at a White House Climate Task Force meeting:

> I considered asking Lindzen if he still believed there was no connection between smoking and lung cancer. He had been a witness for tobacco companies decades earlier, questioning the reliability of statistical connections between smoking and health problems. But I decided that would be too confrontational. When I met him at a later conference, I did ask that question, and was surprised by his response: He began rattling off all the problems with the data relating smoking to health problems, which was closely analogous to his views of climate data. (*Storms of My Grandchildren: The Truth About the Coming Climate Catastrophe and Our Last Chance to Save Humanity* [New York: Bloomsbury Press, 2010], 16)

9. Mann, *Hockey Stick and the Climate Wars*, 69.
10. Recent studies supporting a positive cloud feedback include A. E. Dessler, "A Determination of the Cloud Feedback from Climate Variations over the Past Decade," *Science* 330 (2010): 1523–1527; and S. C. Sherwood, S. Bony, and J. Dufresne, "Spread in Model Climate Sensitivity Traced to Atmospheric Convective Mixing," *Nature* 505 (2014): 37–42.
11. Mann, *Hockey Stick and the Climate Wars*, 69.
12. James Taylor, "New NASA Data Blow Gaping Hole in Global Warming Alarmism" [press release], *Yahoo! News*, July 27, 2011, http://news.yahoo .com/nasa-data-blow-gaping-hold-global-warming-alarmism-192334971.html.
13. The entire affair is described in an editorial published by the resigning editor in chief, Wolfgang Wagner, "Taking Responsibility on Publishing the Controversial Paper 'On the Misdiagnosis of Surface Temperature Feedbacks from Variations in Earth's Radiant Energy Balance' by Spencer and Braswell, *Remote Sens.* 2011, 3(8), 1603–1613," *Remote Sensing* 3 (2011): 2002–2004, doi:10.3390/ rs3092002.
14. U.S. Fish and Wildlife Service, "Polar Bear (*Ursus maritimus*)," ECOS: Environmental Conservation Online System, n.d., http://ecos.fws.gov/species Profile/profile/speciesProfile.action?spcode=A0IJ; International Union for

Conservation of Nature and Natural Resources, "*Ursus maritimus*," IUCN Red List of Threatened Species, 2015, http://www.iucnredlist.org/details/22823/0.

15. Bjorn Lomborg, "Let the Data Speak for Itself [*sic*]," *Guardian*, October 14, 2008, http://www.theguardian.com/commentisfree/2008/oct/14/climatechange -scienceofclimatechange.

16. See the two commentaries by science journalist Greg Laden: "Bjørn Lomborg WSJ Op Ed Is Stunningly Wrong," *ScienceBlogs*, February 3, 2015, http:// scienceblogs.com/gregladen/2015/02/03/bjorn-lomborg-did-not-get-facts -straight/, and "Lomborg in Oz," *ScienceBlogs*, April 23, 2015, http://science blogs.com/gregladen/2015/04/23/lomborg-in-oz/.

17. For the larger pattern of Lomborg's misrepresentations, see Howard Friel, *The Lomborg Deception: Setting the Record Straight About Global Warming* (New Haven, Conn.: Yale University Press, 2011).

18. Such claims abound in the echo chamber of climate change denialism. See, for a representative example, Green Agenda, "What About Greenland?" http:// www.green-agenda.com/greenland.html.

19. Greenland has a surface area of roughly 772,000 square miles (2 million square kilometers), but 16 feet (5 meters) of sea-level rise would lead to the flooding of nearly 1,544,000 square miles (4 million square kilometers) of coastal land area (Michael E. Mann and Lee R. Kump, *Dire Predictions: Understanding Climate Change*, 2nd ed. [New York: DK, 2015], 12–13).

20. See, for example, Roger Pielke Jr., "Disasters Cost More Than Ever—but Not Because of Climate Change," *FiveThirtyEight*, March 19, 2014, http:// fivethirtyeight.com/features/disasters-cost-more-than-ever-but-not-because -of-climate-change/.

21. International Federation of Red Cross and Red Crescent Societies, *World Disasters Report* (Geneva: International Federation of Red Cross and Red Crescent Societies, 2014), http://www.ifrc.org/publications-and-reports /world-disasters-report/world-disasters-report-2014/.

22. Eric Reguly, "No Climate-Change Deniers to Be Found in the Reinsurance Business," *Globe and Mail*, November 28, 2013, http://www .theglobeandmail.com/report-on-business/rob-magazine/an-industry-that -has-woken-up-to-climate-change-no-deniers-at-global-resinsurance-giant /article15635331.

23. David Roberts, "Our Old Friend," *Grist*, January 30, 2007, http://grist.org /article/house-committee-hearings-on-politicization-of-climate-science -guess-who-the/.

24. Kerry Emanuel, "MIT Climate Scientist Responds on Disaster Costs and Climate Change," *FiveThirtyEight*, March 31, 2014, http://fivethirtyeight .com/features/mit-climate-scientist-responds-on-disaster-costs-and-climate -change/; Neville Nicholls, "Comments on 'Have Disaster Losses Increased Due to Anthropogenic Climate Change?'" *Bulletin of the American Meteorological Society* 92 (2011): 791.

25. Quoted in Michael Babad, "Exxon Mobil CEO: 'What Good Is It to Save the Planet If Humanity Suffers?'" *Globe and Mail*, May 30, 2013, http:// www.theglobeandmail.com/report-on-business/top-business-stories /exxon-mobil-ceo-what-good-is-it-to-save-the-planet-if-humanity-suffers /article12258350/.

26. See, for example, Bill Gates, "Two Videos That Illuminate Energy Poverty," *GatesNotes*, June 25, 2014, http://www.gatesnotes.com/Energy/Two-Videos -Illuminate-Energy-Poverty-Bjorn-Lomborg.

27. Paul Thacker, "The Breakthrough Institute's Inconvenient History with Al Gore," Edmond J. Safra Center for Ethics, Harvard University, April 14, 2014, http:// ethics.harvard.edu/blog/breakthrough-institutes-inconvenient-history- al-gore.

28. Among the Breakthrough Institute's primary funders are the Cynthia & George Mitchell Foundation, tied to George Mitchell's fortune derived from natural gas extraction and fracking ("Who Funds Us," http://thebreakthrough.org /about/funders/). The foundation advocates for the continued extraction of natural gas ("Shale Sustainability," Cynthia & George Mitchell Foundation, n.d., http://www.cgmf.org/p/shale-sustainability-program.html).

29. Pope Francis, "Address of His Holiness Pope Francis to the Members of the Diplomatic Corps Accredited to the Holy See," January 13, 2014, English translation provided by the Vatican at http://w2.vatican.va/content/francesco/en /speeches/2014/january/documents/papa-francesco_20140113_corpo- diplomatico.html.

30. R. Jai Krishna, "Renewable Energy Powers Up Rural India," *Wall Street Journal*, July 29, 2015, http://www.wsj.com/articles/renewable-energy-powers -up-rural-india-1438193488.

31. Pope Francis, "Address of His Holiness Pope Francis"; Department of Defense, "DoD Releases Report on Security Implications of Climate Change," July 29, 2015, http://www.defense.gov/news/newsarticle.aspx?id=129366. The Defense Department report notes that "global climate change will aggravate problems such as poverty, social tensions, environmental degradation,

ineffectual leadership and weak political institutions that threaten stability in a number of countries."

32. "World Bank Says Climate Change Could Thrust 100 Million into Deep Poverty by 2030," Fox News, November 8, 2015, http://www.foxnews .com/world/2015/11/08/world-bank-says-climate-change-could-thrust -100-million-into-deep-poverty-by/.

5. The War on Climate Science

1. Jethro Mullen, Yoko Wakatsuki, and Chandrika Narayan, "Hiroo Onoda, Japanese Soldier Who Long Refused to Surrender, Dies at 91," CNN, January 17, 2014, http://www.cnn.com/2014/01/17/world/asia/japan-philippines-ww2-soldier-dies/.

2. Chris Mooney, *The Republican War on Science* (New York: Basic Books, 2005).

3. David Michaels, *Doubt Is Their Product: How Industry's Assault on Science Threatens Your Health* (New York: Oxford University Press, 2008), 4.

4. Naomi Oreskes and Eric M. Conway, *Merchants of Doubt: How a Handful of Scientists Obscured the Truth on Issues from Tobacco Smoke to Global Warming* (New York: Bloomsbury Press, 2010).

5. Ibid., chaps. 1 and 5.

6. Rachel Carson, *Silent Spring* (Boston: Houghton Mifflin, 1962).

7. Quoted in Christopher J. Bosso, *Pesticides and Politics: The Life Cycle of a Public Issue* (Pittsburgh: University of Pittsburgh Press, 1987), 116.

8. Robin McKie, review of *Merchants of Doubt*, by Naomi Oreskes and Eric M. Conway, *Observer*, August 7, 2010, http://www.theguardian.com/books/2010 /aug/08/merchants-of-doubt-oreskes-conway.

9. Competitive Enterprise Institute, "Dangerous Legacy," 2016, http://www .rachelwaswrong.org.

10. G. E. Likens, F. H. Bormann, and N. M. Johnson, "Acid Rain," *Environment* 14 (1972): 33–40; G. E. Likens and F. H. Bormann, "Acid Rain: A Serious Regional Environmental Problem," *Science* 184 (1974): 1176–1179.

11. Oreskes and Conway, *Merchants of Doubt*, chap. 3.

12. Center for Public Integrity, "Stealth PACs Revealed," February 9, 2000, http:// www.publicintegrity.org/2000/02/09/3311/stealth-pacs-revealed; Lee Fang, "Promoting Acid Rain to Climate Denial: Over 20 Years of David Koch's Polluter Front Groups," Center for American Progress, April 1, 2010, http://think progress.org/climate/2010/04/01/174612/koch-pollution-astroturf-2deca/.

13. Quoted in *Chemical Week,* July 16, 1975.
14. Oreskes and Conway, *Merchants of Doubt,* chap. 4.
15. Mooney, *Republican War on Science,* 4.
16. Oreskes and Conway, *Merchants of Doubt,* passim.
17. Sharon Begley, "The Truth About Denial," *Newsweek,* August 13, 2007.
18. Oreskes and Conway, *Merchants of Doubt,* chap. 2.
19. Michael Mann, *The Hockey Stick and the Climate Wars: Dispatches from the Front Lines* (New York: Columbia University Press, 2013), 65.
20. Ibid., 84–87.
21. Robert Proctor, "Manufactured Ignorance," review of *Merchants of Doubt,* by Naomi Oreskes and Eric M. Conway, *American Scientist,* September –October 2010, http://www.americanscientist.org/bookshelf/pub/manufactured -ignorance.
22. For information on Frederick Seitz—covering everything from his own opinions to his connections to the tobacco and oil industries and his role in denying the negative impacts of tobacco and the burning of fossil fuels—see Oreskes and Conway, *Merchants of Doubt,* 5–6, 8–9, 10–11, 25–29, 35, 36, 37, 54, 56, 151, 186–190, 208–210, 213–214, 238, 244–245, 270–271.
23. Quoted in James Hoggan and Richard Littlemore, *Climate Cover-Up: The Crusade to Deny Global Warming* (Vancouver: Greystone Books, 2009), 90.
24. Mann, *Hockey Stick and the Climate Wars,* 66.
25. Ibid.
26. The Science and Environmental Policy Project has received money from Philip Morris, Texaco, and Monsanto (Keith Hammond, "Wingnuts in Sheep's Clothing," *Mother Jones,* December 4, 1997).
27. Richard Littlemore, "The Deniers? The World Renowned Scientist Who Got Al Gore Started," *DeSmogBlog,* April 16, 2008, http://www.desmogblog.com /the-deniers-the-world-renowned-scientist-who-got-al-gore-started.
28. Dan Harris et al., "Global Warming Denier: Fraud or 'Realist'?" ABC News, March 23, 2008, http://abcnews.go.com/Technology/GlobalWarming/story? id=4506059&page=1.
29. Oreskes and Conway, *Merchants of Doubt,* 20.
30. Justin Gillis and John Schwartz, "Deeper Ties to Corporate Cash for Doubtful Climate Researcher," *New York Times,* February 21, 2015.
31. Oreskes and Conway, *Merchants of Doubt,* 245. A "Potemkin village" is a fake village built to fool and impress visitors.

32. Evidence of funding provided by the fossil fuel industry (in particular, Koch Industries and ExxonMobil) for each of these groups is given in Greenpeace USA, "Koch Industries: Secretly Funding the Climate Denial Machine," March 2010, http://www.greenpeace.org/usa/global-warming/climate-deniers /koch-industries/; Ross Gelbspan, *Boiling Point: How Politicians, Big Oil and Coal, Journalists, and Activists Are Fueling the Climate Crisis—and What We Can Do to Avert Disaster* (New York: Basic Books, 2004); Oreskes and Conway, *Merchants of Doubt*; and James Lawrence Powell, *The Inquisition of Climate Science* (New York: Columbia University Press, 2011).

33. Proctor, "Manufactured Ignorance."

34. Powell, *Inquisition of Climate Science*, 57.

35. In *Merchants of Doubt*, Oreskes and Conway show that a number of climate change deniers, including Singer, served as advisers to the Advancement of Sound Science Coalition, a Philip Morris front group that challenged the evidence linking secondhand smoke to disease.

36. Ibid., 150, 151, 238, 247, 253.

37. Mooney, *Republican War on Science*, 67–68.

38. Tom Philpott, "Did Scientists Just Solve the Bee Collapse Mystery?" *Mother Jones*, May 20, 2014, http://www.motherjones.com/tom-philpott/2014/05 /smoking-gun-bee-collapse. See also "Bees Prefer Foods Containing Neonicotinoid Pesticides," *Nature*, May 7, 2015, 74–76.

39. "A Sharp Spike in Honeybee Deaths Deepens a Worrisome Trend," *New York Times*, May 13, 2015, http://www.nytimes.com/2015/05/14/us/honeybees -mysterious-die-off-appears-to-worsen.html?_r=0.

40. Quoted in Sara Jerving, "Syngenta's Paid Third Party Pundits Spin the 'News' on Atrazine," *PR Watch*, Center for Media and Democracy, February 7, 2012, http://www.prwatch.org/news/2012/02/11276/syngentas-paid-third-party -pundits-spin-news-atrazine.

41. Steven J. Milloy, "Freaky-Frog Fraud," Fox News, November 8, 2002, http:// www.foxnews.com/story/2002/11/08/freaky-frog-fraud.html, quoted in Jerving, "Syngenta's Paid Third Party Pundits."

42. Milloy does hold a bachelor's in natural sciences from Johns Hopkins University and a master's from the Johns Hopkins University School of Hygiene and Public Health, but he has no doctorate in science. He has a juris doctor from the University of Baltimore and a master of laws from the Georgetown University Law Center.

43. Mooney, *Republican War on Science*, 67–68.

44. Paul Thacker, "Smoked Out," *New Republic*, February 6, 2006, http://www
.newrepublic.com/article/104858/smoked-out, quoting Fox News.

45. Henry I. Miller, "Rachel Carson's Deadly Fantasies," *Forbes*, September 5,
2012, http://www.forbes.com/sites/henrymiller/2012/09/05/rachel-carsons
-deadly-fantasies/.

46. "Henry I. Miller," *Sourcewatch*, last modified April 23, 2015, http://www
.sourcewatch.org/index.php/Henry_I._Miller.

47. Hammond, "Wingnuts in Sheep's Clothing." See also J. Justin Lancaster,
"The Cosmos Myth," OSS: Open Source Systems, Science, Solutions, updated
July 6, 2006, http://ossfoundation.us/projects/environment/global-warming
/myths/revelle-gore-singer-lindzen; and "A Note About Roger Revelle, Justin
Lancaster, and Fred Singer," *Blogspot*, September 13, 2004, http://www.rabett
.blogspot.com (contains a comment by Justin Lancaster stating his views
about these matters).

48. S. Fred Singer, "Gore's 'Global Warming Mentor,' in His Own Words,"
Heartland Institute, January 1, 2000, http://news.heartland.org/newspaper
-article/2000/01/01/gores-global-warming-mentor-his-own-words.

49. Carolyn Revelle Hufbauer, "Global Warming: What My Father Really Said"
[op-ed], *Washington Post*, September 13, 1992, https://www.washingtonpost
.com/archive/opinions/1992/09/13/global-warming-what-my-father-really
-said/5791977b-74b0-44f8-a40c-c1a5df6f744d/.

50. Quoted in Ed Regis, "The Doomslayer," *Wired*, February 1997.

51. Union of Concerned Scientists, "World Scientists' Warning to Humanity," November 18, 1992, http://www-formal.stanford.edu/jmc/progress/ucs-statement.txt.

52. "A Joint Statement by Fifty-eight of the World's Scientific Academies," in
National Academy of Sciences, *Population Summit of the World's Scientific
Academies* (Washington, D.C.: National Academies Press, 1993), ii, http://
www.nap.edu/openbook.php?record_id=9148&page=R2.

53. Mann, *Hockey Stick and the Climate Wars*, 76.

54. Quoted in Jonathan Schell, "Our Fragile Earth," *Discover*, October 1989
(emphasis added).

55. Julian L. Simon, "Resources and Population: A Wager," *American Physical
Society Newsletter*, March 1996, http://www.aps.org/publications/apsnews
/199603/upload/mar96.pdf, emphasis added.

56. Intergovernmental Panel on Climate Change, *Climate Change 1995: The Science
of Climate Change, Contribution of Working Group I to the Second Assessment
Report of the Intergovernmental Panel on Climate Change*, ed. J. T. Houghton et
al. (Cambridge: Cambridge University Press, 1995), 22.

57. Benjamin D. Santer reported Singer's accusation in *Hearing Before the House Select Committee on Energy Independence and Global Warming*, 111th Cong., 2nd sess., May 20, 2010.

58. Marc Morano, "Kerry 'Unfit to Be Commander-in-Chief,' Say Former Military Colleagues," CNS, May 3, 2004.

59. Marc Morano, "Time for Meds? NASA Scientist James Hansen Endorses Book Which Calls for 'Ridding the World of Industrial Civilization'—Hansen Declares Author 'Has It Right . . . the System Is the Problem,'" *Climate Depot*, January 22, 2010, http://www.climatedepot.com/a/7355/a/4993 /Time-for-Meds-NASA-scientist-James-Hansen-endorses-book-which-calls -for-ridding-the-world-of-Industrial-Civilization-ndash-Hansen-declares -author-has-it-rightthe-system-is-the-problem.

60. Quoted in Clive Hamilton, "Silencing the Scientists: The Rise of Right-Wing Populism," *Our World*, March 2, 2011, http://ourworld.unu.edu/en /silencing-the-scientists-the-rise-of-right-wing-populism/#authordata.

61. The attacks on Mann are recounted at various points in Mann, *Hockey Stick and the Climate Wars*.

6. Hypocrisy—Thy Name Is Climate Change Denial

1. Scott Harper, "Lawmakers Avoid Buzzwords on Climate Change Bills," *Virginia-Pilot*, June 10, 2012, http://hamptonroads.com/2012/06/lawmakers -avoid-buzzwords-climate-change-bills.

2. For a fuller description of this incident, see Michael Mann, *The Hockey Stick and the Climate Wars: Dispatches from the Front Lines* (New York: Columbia University Press, 2013), 84–87. Tom Toles did two editorial cartoons on the matter. The characterization of Cuccinelli's actions as a "witch hunt" comes from "Ken Cuccinelli's Climate-Change Witch Hunt" [editorial], *Washington Post*, March 11, 2012, https://www.washingtonpost.com/opinions/ken -cuccinellis-climate-change-witch-hunt/2012/03/08/gIQApmdu5R_story .html.

3. Mann, *Hockey Stick and the Climate Wars*, 110.

4. Joe Romm, "Ken Cuccinelli's New Business Will Not Survive Climate Change," Center for American Progress," January 6, 2015, http://thinkprogress.org /climate/2015/01/06/3608217/ken-cuccinelli-irony-alert/.

5. "Prosecutors Say McDonnell Should Begin Prison Term," *Daily Press*, August 14, 2015, http://www.dailypress.com/news/dp-ap-prosecutors-oppose-mcdonnell -bid-story.html.

6. Jennifer Weeks and *Daily Climate*, "Whatever You Call It, Sea Level Rises in Virginia," *Scientific American*, August 21, 2012, http://www.scientificamerican.com/article/whatever-you-call-it-sea-level-rises-in-virginia/.

7. Kate Sheppard, "North Carolina Wishes Away Climate Change," *Mother Jones*, June 1, 2012, http://www.motherjones.com/blue-marble/2012/05/north-carolina-wishes-away-climate-change.

8. Brian Merchant, "10 Feet of Global Sea Level Rise Is Now Guaranteed," *Motherboard*, May 12, 2014, http://motherboard.vice.com/read/10-feet-of-global-sea-level-rise-now-inevitable. See also Elizabeth Kolbert, "The Siege of Miami," *New Yorker*, December 21 and 28, 2015, http://www.newyorker.com/magazine/2015/12/21/the-siege-of-miami.

9. Doyle Rice, "Fla. Gov. Bans the Terms 'Climate Change,' 'Global Warming,'" *USA Today*, March 9, 2015, http://www.usatoday.com/story/weather/2015/03/09/florida-governor-climate-change-global-warming/24660287/.

10. Stephen Stromberg, "Rubio's Intellectually Hollow Position on Climate Change," *Washington Post*, April 19, 2015, http://www.washingtonpost.com/blogs/post-partisan/wp/2015/04/19/rubios-intellectually-hollow-position-on-climate-change/.

11. J. Hoggan and R. Littlemore, *Climate Cover-Up: The Crusade to Deny Global Warming* (Vancouver: Greystone Books, 2009), 96.

12. Senator James M. Inhofe (R-Okla.), "Climate Change Update" [floor speech], Senate, 108th Cong., 2nd sess., January 4, 2005.

13. Mann, *Hockey Stick and the Climate Wars*, 117–119.

14. Gavin Schmidt and Michael Mann, "Inhofe and Crichton: Together at Last!" *RealClimate*, September 28, 2015, http://www.realclimate.org/index.php/archives/2005/09/inhofe-and-crichton-together-at-last/.

15. Chris Mooney, *Storm World: Hurricanes, Politics, and the Battle over Global Warming* (New York: Harcourt, 2007), 89.

16. "Gray and Muddy Thinking About Global Warming," *RealClimate*, April 26, 2006, http://www.realclimate.org/index.php/archives/2006/04/gray-on-agw/?wpmp_tp=1.

17. Quoted in Schmidt and Mann, "Inhofe and Crichton."

18. Stephen Lacey, "After Getting Sick from Algae Bloom Exacerbated by Heat Wave and Drought, Inhofe Jokes the 'Environment Strikes Back,'" Center for American Progress, July 1, 2011, http://thinkprogress.org/climate/2011/07/01/259859/algae-bloom-sick-inhofe/; "2011 Texas Drought Was 20 Times More Likely Due to Warming, Study Says," NBC News, July 10, 2012,

http://usnews.nbcnews.com/_news/2012/07/10/12665235-2011-texas
-drought-was-20-times-more-likely-due-to-warming-study-says?lite.

19. "Climate Myths from Joe Barton," *Skeptical Science*, n.d., https://www.skeptical
science.com/skepticquotes.php?s=31.

20. Mann, *Hockey Stick and the Climate Wars*, 160, 164–175, 241–244.

21. Richard Adams, "Joe Barton: The Republican Who Apologised to BP,"
Guardian, June 17, 2010, http://www.theguardian.com/world/richard-adams
-blog/2010/jun/17/joe-barton-bp-apology-oil-spill-republican.

22. Tom Toles, "Ever so Sorry" [cartoon], *Washington Post*, June 22, 2010.

23. U.S. Senate, Committee on Commerce, Science, and Transportation, *Data
or Dogma? Promoting Open Inquiry in the Debate over the Magnitude of
Human Impact on Earth's Climate* [hearing], 114th Cong., 1st sess., December 8,
2015, http://www.commerce.senate.gov/public/index.cfm/hearings?ID=CA2
ABC55-B1E8-4B7A-AF38-34821F6468F7.

24. Tony Dokoupil, "How Climate Change Deniers Got It Right—but Very
Wrong," MSNBC, June 16, 2015, http://www.msnbc.com/msnbc/how-climate
-change-deniers-got-it-very-wrong.

25. Quoted in Samantha Page, "Ted Cruz Invited a Right-Wing Radio Host to
Testify on Climate Science and He Gave This Insane Rant," Center for
American Progress, December 9, 2015, http://thinkprogress.org/climate/2015
/12/09/3729959/mark-steyn-said-some-weird-things-at-this-hearing/.

26. Michael E. Mann, "The Assault on Climate Science," *New York Times*,
December 8, 2015, http://www.nytimes.com/2015/12/08/opinion/the-assault
-on-climate-science.html?.

27. See, for example, Andra J. Reed et al., "Increased Threat of Tropical Cyclones
and Coastal Flooding to New York City During the Anthropogenic Era,"
Proceedings of the National Academy of Sciences 112 (2015): 12610–12615 (and
references therein).

28. Michael E. Mann and Peter H. Gleick, "Climate Change and California
Drought in the 21st Century," *Proceedings of the National Academy of Sciences*
112 (2015): 3858–3859.

29. For a fuller discussion of the manufactured "Climategate" scandal, see Mann,
Hockey Stick and the Climate Wars, chap. 14.

30. Quoted in Lisa Lerer, "Saudi Arabia Calls for 'Climategate' Investigation,"
Politico, December 7, 2009.

31. Sarah Palin, "Sarah Palin on the Politicization of the Copenhagen Climate
Conference," *Washington Post*, December 9, 2009.

32. Brian Montopoli, "Sarah Palin Emails Released from Time as Governor—But Many Withheld or Redacted, CBS News, June 11, 2011, http://www.cbsnews .com/news/sarah-palin-emails-released-from-time-as-governor-but-many -withheld-or-redacted/.

33. Quoted in "Rupert Murdoch Mocks Global Warming with Icy Photo, Enrages Twitter—Again," *Hollywood Reporter*, February 27, 2015, http://www.hollywood reporter.com/news/rupert-murdoch-mocks-global-warming-778302.

34. Quoted in Dana Nuccitelli, "Rupert Murdoch Doesn't Understand Climate Change Basics, and That's a Problem," *Guardian*, July 14, 2014, http://www .theguardian.com/environment/climate-consensus-97-per-cent/2014 /jul/14/rupert-murdoch-doesnt-understand-climate-basics.

35. Xeni Jardin, "Climate Change Denier Rupert Murdoch Just Bought *National Geographic*, Which Gives Grants to Scientists," *Boing Boing*, September 9, 2015, http://boingboing.net/2015/09/09/rupert-murdoch-just-bought-nat .html.

36. Lisa Graves, "The Koch Brothers: The Extremist Roots Run Deep," Center for Media and Democracy, July 10, 2014, http://www.prwatch.org /news/2014/07/12531/koch-brothers-roots-run-deep.

37. Jane Mayer, "Covert Operations: The Billionaire Brothers Who Are Waging a War Against Obama," *New Yorker*, August 30, 2010.

38. Jared Gilmour, "Keystone XL Pipeline Could Yield $100 Billion for Koch Brothers," *Huffington Post*, October 21, 2013, http://www.huffingtonpost .com/2013/10/21/keystone-xl-koch-brothers_n_4136491.html.

39. James Hansen, "Game over for the Climate," *New York Times*, May 9, 2012.

40. Ashley Alman, "Koch Brothers Net Worth Soars Past $100 Billion," *Huffington Post*, April 16, 2014, http://www.huffingtonpost.com/2014/04/16/koch -brothers-net-worth_n_5163010.html.

41. *Citizens United v. Federal Election Commission*, 558 U.S. 310 (2010); Eric Lichtblau, "Advocacy Group Says Justices May Have Conflict in Campaign Finance Cases," *New York Times*, January 21, 2011, http://www.nytimes .com/2011/01/20/us/politics/20koch.html?_r=0.

42. Fredreka Schouten, "Koch Brothers Set $889 Million Budget for 2016," *USA Today*, January 27, 2015, http://www.usatoday.com/story/news /politics/2015/01/26/koch-brothers-network-announces-889-million -budget-for-next-two-years/22363809/.

43. Greenpeace USA, "Koch Industries: Secretly Funding the Climate Denial Machine," March 2010, http://www.greenpeace.org/usa/global-warming /climate-deniers/koch-industries/.

44. Alex Roarty, "The Koch Network Spent $100 Million This Election Cycle," *National Journal*, November 4, 2014, http://www.nationaljournal.com /politics/the-koch-network-spent-100-million-this-election-cycle-20141104.

45. Americans for Prosperity, "Welcome to the Hot Air Tour," http://www .hotairtour.org/ (accessed August 2008).

46. Mann, *Hockey Stick and the Climate Wars*, 216.

47. J. J. Sutherland, "'They Call It Pollution. We Call It Life,'" National Public Radio, May 23, 2006, http://www.npr.org/templates/story/story .php?storyId=5425355.

48. Dan Harris et al., "Global Warming Denier: Fraud or 'Realist'?" ABC News, March 23, 2008, http://abcnews.go.com/Technology/GlobalWarming/story? id=4506059&page=1.

49. Mann, *Hockey Stick and the Climate Wars*, 268.

50. Dara Kerr, "Microsoft Aims to Be Greener and Drops ALEC Membership," CNET, August 19, 2014, http://www.cnet.com/news/microsoft-aims-to-be -greener-and-drops-alec-membership/.

51. Quoted in Suzanne Goldenberg, "Google to Cut Ties with Rightwing Lobby Group over Climate Change 'Lies,'" *Guardian*, September 22, 2014, http://www.theguardian.com/environment/2014/sep/23/google-to-cut -ties-with-rightwing-lobby-group-over-climate-change-lies.

52. John Timmer, "BP Pulls Out of Climate Denial Group Even as Execs Support Sen. Inhofe: ALEC Continues to Lose Corporate Backers over Climate Issues," *Ars Technica*, March 25, 2015.

53. Quoted in Karl Mathiesen and Ed Pilkington, "Royal Dutch Shell Cuts Ties with ALEC over Rightwing Group's Climate Denial," *Guardian*, August 7, 2015.

54. Nicole Brown, "Here Are the 2016 Candidates Who Take Money from the Koch Brothers," *Slant News*, July 15, 2015, https://www.slantnews.com /story/2015-07-15-these-are-the-presidential-candidates-who-do-koch.

55. Loren Gutentag, "Koch Brothers Spread the Wealth Among GOP Candidates," *Newsmax*, October 21, 2015, http://www.newsmax.com/Politics /Koch-Brothers-Jeb-Bush-Marco-Rubio/2015/10/21/id/697273/.

56. Brown, "Here Are the 2016 Candidates Who Take Money from the Koch Brothers."

57. Colin Campbell, "Report: One of the Koch Brothers Just Revealed Which Republican 2016 Candidate They Support," *Business Insider*, April 20, 2015, http://www.businessinsider.com/report-the-koch-brothers-are-backing -scott-walker-2015-4#ixzz3j6xoGsz0.

58. Tim McDonnell, "Scott Walker Is the Worst Candidate for the Environment," *Mother Jones*, March 11, 2015.

59. Neela Banerjee, "Groups Want David Koch Unseated from Smithsonian, AMNH Boards," *Inside Climate News*, March 25, 2015, http://insideclimatenews .org/news/24032015/groups-want-david-koch-unseated-smithsonian-amnh -boards.

60. Natural History Museum, "An Open Letter to Museums from Members of the Scientific Community," March 24, 2015, http://thenaturalhistorymuseum. org/open-letter-to-museums-from-scientists/; Meredith Hoffman, "Leading Scientists Tell the Nation's Museums to Sever Ties with the Koch Brothers," *Vice News*, March 24, 2015, https://news.vice.com/article/leading-scientists- tell-the-nations-museums-to-sever-ties-with-the-koch-brothers.

61. Daniel Souweine, "Why Is WGBH Legitimizing David Koch's Climate Change Denial?" *Huffington Post*, October 15, 2013, http://www.huffingtonpost.com /daniel-souweine/wgbh-david-koch_b_4099534.html.

62. Quoted in Jane Mayer, "A Word from Our Sponsor," *New Yorker*, May 27, 2013, http://www.newyorker.com/magazine/2013/05/27/a-word-from-our -sponsor.

63. Kathleen Miles, "If Koch Brothers Buy *LA Times*, Half of Staff May Quit," *Huffington Post*, April 29, 2013, http://www.huffingtonpost.com/kathleen -miles/koch-brothers-la-times_b_3180391.html.

64. Graham Readfearn, "Is Bjorn Lomborg Right to Say Fossil Fuels Are What Poor Countries Need?" *Guardian*, December 6, 2013, http://www.theguardian .com/environment/planet-oz/2013/dec/06/bjorn-lomborg-climate-change -poor-countries-need-fossil-fuels.

65. Bjorn Lomborg, "Who's Afraid of Climate Change?" Project Syndicate, August 11, 2010, http://www.project-syndicate.org/commentary/who-s-afraid -of-climate-change.

66. Graham Readfearn, "Bjorn Lomborg Think Tank Funder Revealed as Billionaire Republican 'Vulture Capitalist' Paul Singer," Greenpeace, February 9, 2015, http://www.greenpeace.org/usa/bjorn-lomborg-think-tank-funder-revealed -billionaire-republican-vulture-capitalist-paul-singer/.

67. "UWA Cancels Contract for Consensus Centre Involving Controversial Academic Bjorn Lomborg," Australian Broadcasting Corporation, May 8, 2015, http://www.abc.net.au/news/2015-05-08/bjorn-lomborg-uwa-consensus -centre-contract-cancelled/6456708.

68. Bjorn Lomborg, "Geoengineering: A Quick, Clean Fix?" *Time*, November 14, 2010, http://content.time.com/time/magazine/article/0,9171,2030804,00.html;

Colin McInnes, "Time to Embrace Geoengineering," Breakthrough, June 27, 2013, http://thebreakthrough.org/index.php/programs/energy-and-climate /time-to-embrace-geoengineering.

7. Geoengineering, or
"What Could *Possibly* Go Wrong?"

1. Bjorn Lomborg, "Geoengineering: A Quick, Clean Fix?" *Time*, November 14, 2010, http://content.time.com/time/magazine/article/0,9171,2030804,00. html; Colin McInnes, "Time to Embrace Geoengineering," Breakthrough, June 27, 2013, http://thebreakthrough.org/index.php/programs/energy-and -climate/time-to-embrace-geoengineering.
2. For a more technical discussion of the flaws in the various proposed geo-engineering schemes, see A. Robock, "20 Reasons Why Geoengineering May Be a Bad Idea," *Bulletin of the Atomic Scientists* 64 (2008): 14–18, 59, doi:10.2968/064002006.
3. Edward Teller, Roderick Hyde, and Lowell Wood, "Global Warming and Ice Ages: Prospects for Physics-Based Modulation of Global Change" (paper prepared for submittal to the Twenty-Second International Seminar on Planetary Emergencies, Erice, Italy, August 20–23, 1997).
4. "Scientists to Stop Global Warming with 100,000 Square Mile Sun Shade," *Telegraph*, February 26, 2009, http://www.telegraph.co.uk/news/earth/envi-ronment/globalwarming/4839985/Scientists-to-stop-global-warming-with-100000-square-mile-sun-shade.html.
5. R. Angel, "Feasibility of Cooling the Earth with a Cloud of Small Spacecraft near the Inner Lagrange Point (L1)," *Proceedings of the National Academy of Sciences* 46 (2006): 17184–17189, doi:10.1073/pnas.0608163103.
6. Robock, "20 Reasons Why Geoengineering May Be a Bad Idea."
7. Eli Kintisch, "Climate Hacking for Profit: A Good Way to Go Broke," *Fortune*, May 21, 2010, http://archive.fortune.com/2010/05/21/news/economy/geo-engineering.climos.planktos.fortune/index.htm.
8. Gaia Vince, "Sucking CO_2 from the Skies with Artificial Trees," BBC, October 4, 2012, http://www.bbc.com/future/story/20121004-fake-trees-to -clean-the-skies.
9. Daniel Hillel, *The Rivers of Eden: The Struggle for Water and the Quest for Peace in the Middle East* (New York: Oxford University Press, 1994).

8. A Path Forward

1. These numbers come from United Nations, Department of Economic and Social Affairs, "Millennium Development Goals Indicators," 2011, http://mdgs.un.org/unsd/mdg/SeriesDetail.aspx?srid=751.
2. C. McGlade and P. Ekins, "The Geographical Distribution of Fossil Fuels Unused When Limiting Global Warming to 2°C," *Nature*, January 2015, 187–190.
3. "President Obama's Tough, Achievable Climate Plan" [editorial], *New York Times*, August 3, 2015, http://www.nytimes.com/2015/08/04/opinion/president-obamas-tough-achievable-climate-plan.html?_r=0.
4. "Dr. Michael Mann on Climate Change," *Real Time with Bill Maher*, HBO, August 7, 2015, https://www.youtube.com/watch?v=nZ2cCPRS-Q8.
5. Gardiner Harris and Coral Davenport, "E.P.A. Announces New Rules to Cut Methane Emissions," *New York Times*, August 18, 2015.
6. Gregory Korte, "Obama: Keystone Pipeline Bill 'Has Earned My Veto,'" *USA Today*, February 25, 2015, http://www.usatoday.com/story/news/politics/2015/02/24/obama-keystone-veto/23879735/.
7. Quoted in Coral Davenport, "Citing Climate Change, Obama Rejects Construction of Keystone XL Oil Pipeline," *New York Times*, November 6, 2015.
8. The recent work by Florida State University tornado expert James Elsner is described in Jill Elish, "Researchers Develop Model to Correct Tornado Records," *Florida State 24/7*, September 5, 2013, http://news.fsu.edu/More-FSU-News/24-7-News-Archive/2013/September/Researchers-develop-model-to-correct-tornado-records.
9. Neil Bhatiya, "Is China Leading the Way on Cap-and-Trade?" *Week*, September 5, 2014, http://theweek.com/articles/444027/china-leading-way-capandtrade.
10. Lenor Taylor, "US and China Strike Deal on Carbon Cuts in Push for Global Climate Change Pact," *Guardian*, November 12, 2014, http://www.theguardian.com/environment/2014/nov/12/china-and-us-make-carbon-pledge.
11. Jacob Gronholt-Pedersen and David Stanway, "China's Coal Use Falling Faster Than Expected," Reuters, March 26, 2015, http://www.reuters.com/article/2015/03/26/china-coal-idUSL3N0WL32720150326.
12. Chris Mooney, "Why the Global Economy Is Growing, but CO_2 Emissions Aren't," *Washington Post*, March 13, 2015, http://www.washingtonpost.com/news/energy-environment/wp/2015/03/13/for-the-first-time-in-40-years-the-world-economy-grew-but-co2-levels-didnt/.

13. Kenneth Bossong, "US Renewable Electrical Generation Hits 14.3 Percent," *U.S. Energy World*, August 27, 2014, http://www.renewableenergyworld .com/articles/2014/08/us-renewable-electrical-generation-hits-14-3- percent.htm.

14. Chris Mooney, "Obama Visits Nevada, the Center of the Solar Boom, to Talk Clean Energy and Climate Change," *Washington Post*, August 24, 2015, http:// www.washingtonpost.com/news/energy-environment/wp/2015/08/24 /obama-visits-nevada-the-center-of-the-solar-boom-to-talk-clean-energy -and-climate-change/.

15. Bossong, "US Renewable Electrical Generation Hits 14.3 Percent."

16. Ian Clover, "Wholesale Grid Parity for Solar Possible by 2020s, Report Finds," *PV Magazine*, October 7, 2014, http://www.pv-magazine.com/news/details /beitrag/wholesale-grid-parity-for-solar-possible-by-2020s-report-finds _100016708/#axzz3jV8C1pG4.

17. Mooney, "Obama Visits Nevada."

18. Eric Schaal, "Best-Selling Electric Vehicles and Hybrids in 2014," Autos CheatSheet, n.d., http://www.cheatsheet.com/automobiles/10-best-selling -electric-vehicles-and-hybrids-in-2014.html/?a=viewall.

19. See, for example, Michael E. Mann, "The Power of Paris: Climate Challenge Remains, but Now We're on the Right Path," *World Post*, December 13, 2015, http://www.huffingtonpost.com/michael-e-mann/paris-climate-change _b_8799764.html?utm_hp_ref=world.

20. Ned Resnikoff and Amanda Sakuma, "The Largest Climate March in History," MSNBC, September 21, 2014, http://www.msnbc.com/msnbc /largest-climate-march-history-kicks-new-york.

21. A video of DiCaprio giving this speech is available at "Leonardi DiCaprio Speaks at UN Climate Change Summit," *Guardian*, September 23, 2014, http://www.theguardian.com/environment/video/2014/sep/23/leonardo -dicaprio-un-climate-change-summit-speech-video.

22. Suzanne Goldenberg, "Heirs to Rockefeller Oil Fortune Divest from Fossil Fuels over Climate Change," *Guardian*, September 22, 2014, http://www .theguardian.com/environment/2014/sep/22/rockefeller-heirs-divest-fossil -fuels-climate-change.

23. James Hansen, *Storms of My Grandchildren: The Truth About the Coming Climate Catastrophe and Our Last Chance to Save Humanity* (New York: Bloomsbury Press, 2010), 10, 164–166, 180, 230 (for the 350-ppm limit), xi, 117, 118 (for the current level).

24. Bill McKibben, "The Case for Fossil-Fuel Divestment," *Rolling Stone*, February 22, 2013, http://www.rollingstone.com/politics/news/the-case-for -fossil-fuel-divestment-20130222.

25. Jason M. Breslow, "Investigation Finds Exxon Ignored Its Own Early Climate Change Warnings," *Frontline*, PBS, September 16, 2015, http://www.pbs .org/wgbh/pages/frontline/environment/investigation-finds-exxon-ignored -its-own-early-climate-change-warnings/.

26. John M. Broder, "Cigarette Makers in a $368 Billion Accord to Curb Lawsuits and Curtail Marketing," *New York Times*, June 21, 1997.

27. Bob Simison, "New York Attorney General Subpoenas Exxon on Climate Research," *InsideClimate News*, November 5, 2015, http://insideclimatenews .org/news/05112015/new-york-attorney-general-eric-schneiderman-subpoena -Exxon-climate-documents.

28. Sherwood Boehlert, "Can the Party of Reagan Accept the Science of Climate Change?" *Washington Post*, November 19, 2010, http://www.washingtonpost .com/wp-dyn/content/article/2010/11/18/AR2010111805451.html.

29. "Former U.S. Congressman Bob Inglis to Receive JFK Profile in Courage Award for Stance on Climate Change," John F. Kennedy Presidential Library and Museum, April 13, 2015, http://www.jfklibrary.org/About-Us/News-and -Press/Press-Releases/2015-Profile-in-Courage-Announcement.aspx.

30. William D. Ruckelshaus et al., "A Republican Case for Climate Action" [op-ed], *New York Times*, August 1, 2013, http://www.nytimes.com/2013/08/02/opinion /a-republican-case-for-climate-action.html?_r=0.

31. George P. Shultz, "A Reagan Approach to Climate Change" [op-ed], *Washington Post*, March 13, 2015, https://www.washingtonpost.com/opinions/a-reagan -model-on-climate-change/2015/03/13/4f4182e2-c6a8-11e4-b2a1-bed1aaea2816 _story.html.

32. Quoted in Heather Rogers, "Current Thinking," *New York Times*, June 3, 2007, http://www.nytimes.com/2007/06/03/magazine/03wwln-essay-t.html? _r=0.

33. Quoted in John Aziz, "Here Comes the Sun," *Week*, June 17, 2014, http://theweek.com/speedreads/574172/new-star-wars-force-awakens -teaser-hints-epic-lightsaber-battle.

INDEX

Numbers in italics refer to pages on which illustrations appear.

Bush, George W., 110, 113
Bush, Jeb, 111

California: disruption of food production/wine industry in, 33, 47; drought in, 18, 21, 33, 35, 47–48, 100, 102–103; and Pacific Coast Action Plan on Climate and Energy, 137; solar power in, 140
Canada, 33, 37. *See also* Keystone XL pipeline
cap-and-trade system, 71, 72
carbon dioxide (CO_2): ancient high levels of, 57–58; carbon capture and sequestration, 132; carbon-cycle feedback, 59; carbon footprint, 132; and "direct air capture" schemes, 125–127; as greenhouse gas, 10–13, 15, 16, 17; and ocean acidification, 16, 34, 42, 43, 123–125; purported benefits of increased levels of, 62; rapid rate of increase of levels of, 11, 43, 58; as source of emissions in U.S., 132; and weed growth, 42
Carson, Rachel, 70, 82, 110
Cato Institute, 78, 84
CEI. *See* Competitive Enterprise Institute
Chicago, heat wave in, 41
Chicago Tribune, 113
China, 36, 39, 132, *138*, 139–140
chlorofluorocarbons (CFCs), 71–72, 75
Christie, Chris, 91, 111, 137
Christy, John, 54, 78
cigarette smoke, 75, 79, 158n.8. *See also* tobacco industry
Citizens for a Sound Economy, 71, 78

Citizens United v. Federal Election Commission, 109, 111
Clean Air Act, 71, 100
Clean Power Plan, 133–135
Clean Water Act, 100
climate change: denial of (*see* climate change denial, strategies of; climate change deniers); economics of inaction on, 13, 45–48, 64–67, *65*; grassroots action on, 136–137; health effects of, 40–42, 47; impacts of, on ecosystems and food webs, 21, 23, 30, 42–45, *43*; impacts of, on human civilization, 31–51, 65–67, 160n.31 (*see also* agriculture; economics; food production; global conflict; national security; water supply); impacts of, on oceans, 16, 19–23 (*see also* oceans; sea-level rise); impacts of, on weather, 18–19, *19*, 24–27, *26* (*see also* drought; flooding; heat waves; hurricanes; rainfall; snowfall; tornadoes; weather); mortality rate from causes related to, 40–42; policy responses to (*see* public policy); and precautionary principle, 12–13; public understanding of (*see* public understanding of science); recommendations for action on, 146–150; science of (*see* climate science); as threat multiplier, 32; tipping points for, 27–29, *28*, *30*
Climate Change Conference (Copenhagen, 2009), 103, 105, 107
Climate Change Conference (Paris, 2015), 140–142, *141*, 145
climate change denial, strategies of: accusations of scientists' ulterior motives, 3–4; arguments about

"exploiting a tragedy," 102; "because we don't know everything, we know nothing" fallacy, 8–10, 25–26, *26*; and cherry-picking data, 9–10, 54–55; *climate change* and *sea-level rise* as "liberal code words," 92; climate scientists misquoted, 86, 104, 105; deliberate misinformation and confusion campaigns, 3, 7–10, 15–16, 53–56, 63–64, 77, 83–84, 86, 96–97, 103–105; denial of existence of climate change, 53–56, 67; denial of ill effects from climate change, 62–64, 67; denial of link between climate change and severe weather, 100–103; denial that climate change is a grave threat, 53, 58–62, 67, 113–115; denial that climate change is caused by humans, 53, 56–58; denial that high cost of action is justified, 64–67, *65*; "energy poverty" argument, 65–67, 113; fraudulent and misleading publications, 76–77, 83–84, 86, 110; historical perspective on, 69–72; lawsuits aimed at obstructing climate science, 93–94; manufactured scandals, 103–105, 110; and "pause" in global warming, 9–10, 54–56, *55*; personal attacks and smear campaigns, 81–89, 103–105; and politicians, 9–10, 91–100; proposed self-correcting nature of climate system, 58–62, *60*; and pseudoskepticism, 1–5; stages of denial, *52*, 53–67 (*see also specific stages*); states' attempts to legislate discussion of climate change,

92–95; "tobacco strategy," 9, 70–72; use of alternative "facts," 7
climate change deniers, *68*; belief of, in technological fixes, 64, 67, 115, 116, 117–129 (*see also* geoengineering); Cold War–era physicists as, 71, 74–78, 82; front organizations and think tanks as, 72–73, *73*, 77, 78, 108–111 (*see also specific organizations*); hypocrisy of, 91–115, 118; ideological backgrounds of, 75, 108; industry executives as, 65–66; industry-funded scientists and lobbyists as, 61, 74–81, *79*, 84, 86–88; Koch brothers–supported front groups, organizations, political candidates, and "experts" as, 108–114, 139; politicians as, 9–10, 91–100, 111–112, 139; previous affiliations of, with tobacco industry, 77, 79–80; previous attacks of, on environmental causes, 70–72, 74–75, 80–82, 98, 158n.8; researchers as, 54, 58–59, 61, 78; Saudi royal family as, 107; science-fiction writers as, 97; self-styled "skeptics" and "experts" as, 62, 64, 66, 113–115; statements by, 62, 66, 105, 106–107, 110. *See also* funding for denial advocacy
"Climategate," 103–105, 110
climate science, 10–13, 15–18, *17*; on equilibrium climate sensitivity, 17; and feedback loops, 11, 17, 58–62; on hockey stick curve, 5–6, 57, 152n.8; and human causation, 16–18; overwhelming scientific consensus on, 9; on "pause" in global warming, 9–10, 54–56, *55*;

energy: Clean Power Plan, 133–135; and costs of delaying action on climate change, 46; "energy poverty" argument, 65–67, 113; food–water–energy nexus, 36–38; from fossil fuels, need to phase out, 132, 134–135; renewable sources of, 118, 134, 140–141, 147–148

Energy and Enterprise Initiative, 145

environmental movement: and fire on Cuyahoga River, 100; industry attacks on, 70–72, 82; and legislation under Republican administrations, 71, 72, 145–146; need for whole-planet environmentalism, 149, 149–150; as target of climate change deniers, 70–72, 74–75, 80–82, 98, 158n.8

Environmental Protection Agency (EPA), 71, 133, 134

environmental refugeeism, 40

equilibrium climate sensitivity, 17

ethics, 49–51

Ethiopia, 47

Europe, heat waves in, 18, 41

extinction of species, 30, 42–45, 43

"extraordinary claims require extraordinary evidence" statement, 1, 4

ExxonMobil, 65, 75, 81, 95, 142, 143

feedback loops, 11, 17, 58–62

flooding, 18, 19, 24, 27, 92, 159n.19; deaths from, 41–42; and high food prices, 48; and politicians' denial of climate change, 92; storm surges as cause of, 38–39. *See also* rainfall

Florida, 20, 95

food production, 32–34; disruption of, 33, 62, 98; and droughts and floods, 47–48, 98; food–water–energy nexus, 36–38; and malnutrition, 40

food webs, 21, 45

fossil fuel industry: climate change deniers funded or employed by, 65–66, 75, 78, 79, 95, 109–111 (*see also* funding for denial advocacy); divestment from, 143; "energy poverty" argument of, 65–67, 113; fuel reserves of, 143; need for phasing out, 132, 134–135; "They Call It Pollution. We Call It Life" campaign of, 110. *See also* Koch, Charles and David; *specific companies*

Fourier, James, 16

Fox News, 7, 80–81, 88, 104, 106, 107

fracking, 37, 45–46, 135

Francis (pope), 49, 50, 51, 66–67, 144, 144

Fraser Institute, 78

FreedomWorks, 71, 78

frogs, 45, 80

front organizations and think tanks, 73, 77; fraudulent publications of, 77; funding for, from Koch brothers, 109–111, 114; funding by, for climate change–denial advocacy, 72–73, 73, 160n.28; listed, 78; and manufactured scandals, 104, 110; and personal attacks against climate scientists, 82, 84, 86, 88, 110; and previous environmental causes, 71. *See also* Koch, Charles and David; *specific organizations*

funding for denial advocacy: and *Citizens United*, 109, 111; by front organizations and think tanks, 72–73, 73, 160n.28; and industry-funded scientists and "experts,"

oysters, 34, 42
ozone layer, depletion of, 10, 71–72, 75, 80, 123

Pacific Coast Action Plan on Climate and Energy, 137
Pakistan, 18, 41, 48
Palin, Sarah, 105, *105*
Paul, Rand, 111
peer-review process, 2–3, 62
penguin, *44*
People's Climate March, 142
permafrost, 11, 59
Perry, Rick, 99, 111
Philip Morris, 77, 79, 80, 81
Philippines, 39
Pielke, Roger, Jr., 64
pika, *44*
Planktos, 125
polar bear, 21, 44, *44*, 62
politicians: and *Citizens United*, 109, 111; denial of climate change by, 10, 91–100, 111–112, 145, 148–149; funding of, by fossil fuel interests, 95, 109, 111–112; and Koch brothers, 109, 111–112, 139, 145; and manufactured scandals, 104–105. *See also* Republican Party
population, growth of, 32–33
Population Bomb, The (Ehrlich), 84
post hoc, ergo propter hoc fallacy, 56
poverty: "energy poverty" argument, 65–67, 113
precautionary principle, *11*, 12–13
Proctor, Robert, 75
public policy, 12–13, *130*; and costs of delaying action on climate change, 13, 45–48, *46*; and global cooperation, *138*, 138–142; need

for executive action on climate change, 133–136; need for reduced emissions, 131–132; and politicians' denial of climate change (*see* politicians); and politics of change, 142–146; and precautionary principle, *11*, 12–13; state-level and grassroots action on climate change, 136–137; states' attempts to legislate discussion of climate change, 92–95; and weakness in public understanding of science, 3
public understanding of science, *14*; and deliberate misinformation and confusion campaigns, 3, 7–10, 15–16, 53–56, 63–64, 77, 83–84, 86, 96–97; and false "balance" promoted by mainstream media, 9, 105–108; and Pope Francis, 50, 144; and pseudoskepticism, 3; and recognition of climate change threat by U.S. military, insurance industry, and business leaders, 144; and recommendations for action on climate change, 146–150; and seepage effect, 55; and severe weather, 100–103; and teachable moments, 100–103. *See also* media

Qatar, 132

Rahmstorf, Stefan, 152n.8
rainfall, 18, *19*, 24, 27, 48. *See also* flooding
Reagan, Ronald, 72, 145
Real Time with Bill Maher, 133
Red Cross, 64
Regional Greenhouse Gas Initiative, 137

solar power, 140
Solomon, Susan, 105
Somalia, 47
Soon, Willie, 76, 78, 95
space exploration, 149–150
special interests. *See* fossil fuel industry; front organizations and think tanks; funding for denial advocacy; Koch, Charles and David; *specific companies*
Spencer, Roy, 54, 61, 78
State of Fear (Crichton), 97
storm surges, 38–39
Strategic Defense Initiative ("Star Wars"), 72–73
sulfate, injections of, into atmosphere, 121–124
Superstorm Sandy, 21, 27, 39, 47, 91–92, 101–102
Supreme Court, U.S., 109
sustainability, 137, 149
Swiss Re, 64
Syngenta, 80
Syria, 32

Taylor, James, 61
technological fixes for climate change. *See* geoengineering
Teller, Edward, 120
temperature. *See* global warming
Texaco, 77
Texas: climate change denial by politicians in, 99; effects of climate change in, 18, 98, 99, 100; wind farms in, 140
Thacker, Paul, 81
Thailand, 39, 48
think tanks. *See* front organizations and think tanks

Thomas, Clarence, 109
Tillerson, Rex, 65–66
tipping points for climate change, 27–29, *28*
tobacco industry: and climate change deniers, 77, 79–80; "doubt is our product" memo in, 9; eventual accountability of, 143; public disinformation campaign by, 6, 9, 69–72. *See also* cigarette smoke
Toles, Tom, 94, 98
Tornado Alley, 24
tornadoes, 24, 92, 136
transportation, emissions from, 133
trees, artificial, 125–126
Tribune Company, 112–113
Trump, Donald, 111, 112, *148*, 148–149
Tuvalu, 20, 128
Tyson, Neil deGrasse, 106

Ukraine, 33, 48
uncertainty, *14*; "because we don't know everything, we know nothing" fallacy, 8–10, 25–26, *26*; and costs of delaying action on climate change, 48; and deliberate misinformation campaigns, 8–10 (*see also* climate change denial, strategies of); and "doubt is our product" memo, 9
Union of Concerned Scientists, 148
United States: carbon footprint of, 132; changes in weather patterns in, 18, 19, *19*, 21, 24, 40; climate change–related costs in, 46–48; droughts in, 18, 21, 33, 35, 47, 100; environmental legislation under Republican administrations in, 71, 72, 145–146; flooding in, 18, 92

United States (*continued*)
(*see also* hurricanes); heat waves in, 18, 40–41; need for executive action on climate change in, 133–136; politicians' denial of climate change in, 91–100, 139, 145; recognition of climate change threat by military, insurance industry, and business leaders in, 144; renewable energy in, 140–141; and sea-level rise, 20, 39; sources of emissions in, 133; state-level and grassroots action on climate change in, 136–137; states' attempts to legislate discussion of climate change in, 92–95; water shortages in, 35. *See also specific states*
University of Virginia, lawsuit against, 93, 93–94
USA Today, 113

Vermont, 137
Vietnam, 39, 48
Virginia, 92–94
volcanoes, 71, 121–122

Walker, Scott, 112
Wall Street Journal, 7, 66, 88, 104, 106, 113
walrus, 21

warfare, 32, 46, 128
"war on science," 6; historical perspective on, 69–72
Washington Post, 94, 105, 146
Washington State, 137
Washington Times, 108
water supply, 34–36; food–water–energy nexus, 36–38; and fracking, 45–46; and mortality rate, 40
Watts, Anthony, 157n.3
weather, 18–19, 24–27, 26, 92; costs of storm damage, 46–47; deaths from severe, 40–42; denial of link between climate change and severe, 100–103; and deniers' arguments about "exploiting a tragedy," 102; and food production, 32–34, 47–48; and teachable moments, 100–103; and water supply, 35. *See also* drought; flooding; heat waves; hurricanes; rainfall; snowfall; tornadoes
West Antarctic Ice Sheet, 20, 29, 122
wildfires, 33, 34, 41–42, 48, 92
wind shear, 25
wine industry, 47

Years of Living Dangerously (television series), 146
Younger Dryas, 23